Study Guide for Jaccard & Becker's

STATISTICS

FOR THE BEHAVIORAL SCIENCES

Third Edition

James Jaccard
State University of New York at Albany

Brooks/Cole Publishing Company

I(T)P® An International Thomson Publishing Company

Pacific Grove • Albany • Belmont • Bonn • Boston • Cincinnati • Detroit
Johannesburg • London • Madrid • Melbourne • Mexico City
New York • Paris • Singapore • Tokyo • Toronto • Washington

Assistant Editor: *Faith B. Stoddard*
Cover Design: *Cheryl Carrington*
Cover Photo: *Pedro Lobo/Photonica, Inc.*
Editorial Assistant: *Nancy Conti*
Marketing Team: *Gay Meixel and Romy Taormina*
Production Editor: *Mary Vezilich*
Printing and Binding: *Edwards Brothers Incorporated*

 The ITP logo is a registered trademark under license.

For more information, contact:

BROOKS/COLE PUBLISHING COMPANY
511 Forest Lodge Road
Pacific Grove, CA 93950
USA

International Thomson Editores
Seneca 53
Col. Polanco
México, D. F., México C. P. 11560

International Thomson Publishing Europe
Berkshire House 168-173
High Holborn
London WC1V 7AA
England

International Thomson Publishing Japan
Hirakawacho Kyowa Building, 3F
2-2-1 Hirakawacho
Chiyoda-ku, Tokyo 102
Japan

Thomas Nelson Australia
102 Dodds Street
South Melbourne, 3205
Victoria, Australia

International Thomson Publishing Asia
221 Henderson Road
#05-10 Henderson Building
Singapore 0315

Nelson Canada
1120 Birchmount Road
Scarborough, Ontario
Canada M1K 5G4

International Thomson Publishing GmbH
Königswinterer Strasse 418
53227 Bonn
Germany

Printed in the United States of America

5   4   3   2   1

ISBN 0-534-17407-8

Contents

Preface

This study guide is intended to help you learn the material in the textbook and to practice and reinforce important concepts. Each chapter has the same basic structure. First, we highlight some of the major study objectives that you should have as you approach the material in the chapter. This section emphasizes the major concepts that you should be sure to understand and have an intuitive feel for. Second, we provide you with some study tips that will help you learn the material and which we have found to be useful to keep in mind. Third, we provide you with a glossary of important terms and their definitions. You should learn these definitions. Finally, we provide you with practice questions that you can use to test your knowledge. These questions have a "true-false" format and a "short answer" format. Answers to all the questions are given at the end of the question section. If you answer any question incorrectly, try to figure out why you missed it. Look at the material in the textbook for the right answer. If you are diligent about working through your answers and figuring out why you gave a wrong answer, then this will help you master the material.

We have tried to structure this study guide so that it is not redundant with the text book or the exercises within the textbook. Rather, we try to help you organize the material and test yourself in slightly different ways than what the textbook exercises do.

Good luck and stay with it! Statistics really isn't that difficult if you apply yourself and give yourself plenty of time for study.

In developing this study guide, I used a structure for presenting answers to selected exercises developed in the study guide for the second edition by Lana Ruddy. James Cranford contributed greatly to the development of practice questions. I extend my sincere gratitude to both of these individuals for their efforts.

Overview

In the chapters that follow, we provide you with material for learning the contents of individual chapters in the main textbook. In this section, we discuss general strategies that you can use to help you learn statistics..

Strategy 1: Study the Material Regularly

One of the biggest mistakes that students in statistics classes make is to not study on a weekly basis. Statistics is not like some courses, where you can "pull an all nighter" and cram six or seven weeks of material into a night or two of intense study. It just won't work. If you have ever taken a foreign language course, you know that such a strategy is ineffective for learning a language. The same is true of statistics. Later material builds on early material and you must keep up.

Set aside *at least* one evening or afternoon a week to go over the material in the text and your lecture notes. Better yet, do this after each class. Structure a quiet time where you will not be interrupted and where you can concentrate. If you do this after each class, then you can ask your instructor questions during the next class, so as to clarify things you may not be clear about. If you wait too long, you might find yourself behind and lost. Stay current!

Strategy 2: Don't be Afraid to Ask the Instructor

If something is not clear to you, ask the instructor to clarify it. If you are embarrassed to ask it in class, then ask your question either before or after class. Or, go see the instructor during his or her office hours. Although many professors may seem unapproachable, the vast majority are dedicated teachers who love to see a motivated student who is working hard to master the material. Professors greatly respect students who try hard and make an extra effort to learn material. Do not be afraid to show ignorance. Having taught statistics for 20 years, we can assure you that instructors have heard a wide array of questions, ranging from mundane and simplistic to intriguing and provocative. Your task is not to impress the instructor one way or the other. Your task is to learn the material. If you do so, you will gain the instructor's respect.

When you ask a question, try to be specific and direct. Questions like "I don't understand Chapter 3; could you explain it to me?" are too vague and general. Be focused in your questions. Try to know what you don't know so that you can ask questions that will be pointed and most informative for you.

Strategy 3: Do all the Exercises in the Book and Study Guide

Do every exercise in the textbook and every exercise in this study guide. In the textbook, we provide the answers to about half of the exercises. We do not provide answers to all of the exercises, because some instructors like to use these for graded homework. Even if your instructor does not assign an exercise, do it. Even if you do not have the answer to an exercise, do it. Sometimes just the act of doing an exercise will reveal that you do not know something you thought you did.

If you get an exercise wrong, try to diagnose why. Was it because of a simple calculational error? Or was it the result of not understanding something. Go back to the textbook and find where the material pertaining to the question is discussed. Re-read it and think it through. Make sure you understand why you missed the problem. If you are not sure, then ask your instructor.

Strategy 4: Use your Calculator Wisely

We recommend that you use a calculator. However, you do not need a fancy one. Do not expect to pay a lot of money for a fancy statistical calculator and think this will give you an edge in the course. It will not. Ultimately, it is *you* who must learn the material, not the calculator. Statistics is much more than a mass of algebraic manipulations. Your instructor wants you to learn statistical *concepts*, not (just) calculational strategies.

We recommend you use your calculator as a way of double checking your calculations. Whenever possible, do the calculations by hand and then double check your result with the calculator.

Strategy 5: Re-write your Class Notes

During class, you take notes quickly to keep up with the instructor. After each class, re-write your notes clearly, elaborating on the "shorthand" you may have been using during class. By re-writing your notes, this helps you to go over the material and learn it. It also can reveal places in your notes that are too sketchy or incomplete or where you may not have really understood a concept. Having clearly written and organized notes will also make it easier for you to study for your exams.

Strategy 6: Read the Book

Do not rely solely on the book and do not rely solely on lectures. Your instructor may present things slightly differently than in the book. Maybe you will understand the way the book presents the material better than the way your instructor does. Or maybe you will understand your instructor better than the book. If you carefully attend to both, it is more likely you will be able to relate to one or the other form of presentation.

We have found the following strategy to be useful. First, read the chapter *before* going to the lecture on that material. Then, after the lecture, re-read the chapter. The first reading helps to familiarize you with the material before the lecture. The second reading helps to reinforce what you have learned and to highlight possible points of confusion.

Strategy 7: Get a Study Partner

If possible, find someone else who is taking the class who you can become a study partner with. Give each other weekly quizzes. Compare your notes with your partner's notes. Are his or her notes more thorough than your notes? Discuss with your partner possible points of confusion. You can learn as much by teaching your partner about a concept that he or she might not understand as you can from having your partner teach you. Meet with your partner on a regular basis.

Strategy 8: Attend all Classes

Sometimes students miss classes because they are ill or have something unavoidable come up that they can do nothing about. However, sometimes students miss a class because they would rather go to lunch with a friend, attend a rally, get an early start to a vacation, or enjoy the weather outside. If at all possible, make sure you attend every lecture you can. Lectures not only expose you to and help you learn material, but they also provide you with information about what concepts the professor thinks are most important. They help you focus on what material you should make a special effort to understand. In some cases, the professor teaches the material in a way that is slightly different from the text, in order to deepen your understanding of the concepts. It is important to get his or her perspective. Poor attendance in a statistics class almost guarantees poor performance.

Strategy 9: Read all of the Material in Boxes and Highlights

The textbook contains a great deal of material in "boxes" or "highlighted" sections. These include "Applications to the Analysis of a Social Problem" and "Study Exercises." Students often view such material as ancillary, because most instructors do not test the student on such material. Thus, there is a tendency to skip these sections, often not even reading them. This is a mistake. These sections reinforce calculational steps and help you integrate what you are reading into a "big picture" view of why social scientists use statistics. If you read these highlighted sections, the statistics will take on more meaning as you see them actually applied to important questions. They help you understand the steps you must go through as you approach data analysis. Be sure to read and think about the highlighted sections.

Strategy 10: Use Multiple Memorization Strategies and Multiple Strategies to Test your Memory

When it is necessary for you to memorize materials, use different strategies to commit material to memory. Some students feel that reading a textbook and underlining or highlighting with a marker the "essential" material is a good way to memorize material. Our experience is that this is not an effective strategy. Memorizing something usually requires going over it many times in your mind. A casual act of highlighting a sentence will not do the trick. Sometimes actively writing out the main points in your own words works. This usually forces you to perform three rehearsals of the material (1) reading it the first time to determine what to write, (2) writing it down, and (3) reading over what you have written to make sure it makes sense. Subvocally rehearsing a definition over and over also can be useful. Another strategy is to try to link the material to an image or something meaningful and easily remembered in your life.

Test yourself on your memory. Try recalling the material in different ways. For example, if you are trying to commit a set of definitions to memory, test yourself by writing each definitional term down on a piece of paper. Then, at a later time, read the list of terms, defining each one aloud. Reverse this strategy. Write down the definitions on a piece of paper. Then, at a later time, read each definition and state aloud the term that goes with it. Test yourself over and over until you achieve perfect recall.

Strategy 11: Role Play being the Instructor

For a given chapter, pretend that you are the instructor and that you need to write an exam that will test students on the material for the chapter. What questions would you ask? Write them down and then write out the answers. Be sure to test the students on all of the major concepts within the chapter. We have found that when students do this type of "role playing," they are often quite accurate at anticipating the questions of the professor and it is an effective way of learning material, if you are conscientious about writing out all of the answers.

Strategy 12: Diagnose your Test Mistakes

After an exam or quiz, make sure you review the results to see what questions you have missed. Determine why you missed the questions. Sometimes a student misses a question because of careless test taking behavior. That is, the student knew the material but, for some reason, simply wrote down the wrong answer during the test. Was this why you missed the question or was it because you truly did not know the concept? If the problem was carelessness during the test, then you need to think about how to take the test more carefully. For example, you might read a question twice before answering it. You might re-read your answers to make sure they say what you want. You might force yourself to pause for a second or two after reading a question in order to collect your thoughts before answering it.

If you simply did not know the material, then try to determine why you did not know it. Did you not know it because you didn't study the material? If so, then you need to change your study habits (by using the strategies discussed above). Or was it the case that you studied the material, but you just misunderstood it? If this is the case, then you need to talk to the instructor to have him or her clarify your misunderstandings. Perhaps you are not taking complete enough lecture notes and you left out important material in your note taking. Compare your notes with those of someone who did well on the test and see if your notes are less complete.

In short, you should not accept your exam performance passively. You want to learn from your mistakes. Take an active role in diagnosing why you missed the questions you did.

Getting a Tutor

Some students decide to get tutoring for their statistics class. A tutor provides you with someone who can answer your questions and help you organize material. You may want a tutor because you feel more comfortable asking the tutor questions than your professor, or because you want to spend a great deal of time working with the tutor, more time than you can reasonably expect to get working with your professor.

One of the best ways to get a tutor is to ask your instructor if he or she knows someone who you can hire. Each professor emphasizes slightly different material when he or she teaches statistics, so it is good to have a tutor that is familiar with your instructor and how s/he teaches the class. Your professor will probably know someone with these qualifications. If not, ask the professor if s/he can tell you the names of the best students in the class when s/he last taught the course. Then contact these individuals to see if they might be willing to be a tutor.

If you hire a tutor, do not think that this guarantees success. After all, you are the one who must learn statistics and the tutor is only a resource to help you do this.

Mastering the Material

If you follow the 12 strategies outlined above and if you make sure that you can accomplish all of the study objectives listed at the beginning of each chapter in this study guide, then we are certain that you will do extremely well in your statistics course. Good luck!

Chapter 1: Introduction and Mathematical Preliminaries

Study Objectives

This chapter introduces a variety of concepts that are the basis for discussion in later chapters. After reading the material, you should be able to describe the general activities involved in conducting research. You should be able to define a variable, a constant, an independent variable, and a dependent variable. You should be able to identify the independent and dependent variable in a study that you read about.

You should be able to define measurement and characterize nominal, ordinal, interval, and ratio level measures. You should be able to define qualitative and quantitative variables as well as discrete and continuous variables. You should be able to identify if a measure is qualitative or quantitative.

You should be able to define the concept of probability and characterize it in terms of behavior "over the long run."

You should be able to define a population, a sample, a parameter, and a statistic. You should be able to explain what a representative sample is as well as the process of random sampling. You should know how to use a table of random numbers.

You should be able to characterize and explain the difference between descriptive statistics and inferential statistics.

You should know summation notation, both in its shorthand form and in its full form. You should know how to define the real limits of a number. You should know the rules for rounding numbers off.

Study Tips

There are a large number of new terms that you probably have not seen in the past. In some ways, statistics is like a foreign language where you have to learn a whole new vocabulary. Study the terms and definitions carefully in this chapter. You must become "fluent" with them. The section on the "glossary of important terms" is particularly important in this chapter. Know your definitions! Be able to state them in your own words.

Summation notation is used throughout the book. Make a special effort to understand this notation scheme, because you will have difficulty unless you master it.

The most common errors that students make as they approach the material in this chapter are with summation notation. There is a tendency to confuse the different forms of notation. For example, ΣX^2 is not the same as $(\Sigma X)^2$, but students sometimes think they are equivalent. Pay particular attention to the different forms of summation notation!

Another area of difficulty is with identifying whether a measure is ordinal, interval, or ratio level. Measures that appear to be ratio at face value, often are not. We have found that students are quick to attribute ratio level properties when it may not be justified. We think of a measure that an investigator obtains as an *indicator* of some construct. For example, a response to the question "Are you male or female?" is an indicator of someone's gender. How quickly someone's heart is beating might be used as an indicator of stress or arousal. How many friends someone says he or she has might be used as an indicator of how popular the person is. A key issue in measurement is how the indicator maps onto the underlying construct. Even though a measure may seem to have ratio properties (e.g., 120 heart beats are twice as many as 60 heart beats), this does not mean that the measure, *as an indicator of the underlying construct*, has ratio properties (i.e., someone with 120 heart beats is twice as stressed as someone with 60 heart beats). It is important to keep this distinction in mind.

Glossary of Important Terms

Study the terms listed below carefully. Make sure you understand each so that you could explain them to someone else who does not know them.

Hypothesis: A statement proposing that something is true about a given phenomenon.

Variables: A phenomenon that takes on different values, or levels.

Constant: Something that does not vary within given constraints.

Independent Variable: In a causal relationship between two variables, the independent variable is the presumed cause of the other variable. It is the variable that is "doing the influencing."

Dependent Variable: In a causal relationship between two variables, the dependent variable is the presumed effect of the other variable. It is the variable that is "being influenced."

Measurement: The process of translating empirical relationships between objects into numerical relationships.

Nominal Measurement: Type of measurement in which numbers are used merely as labels.

Ordinal Measurement: Type of measurement in which the assignment of numbers permit one to order objects or people along a continuum or dimension.

Interval Measurement: Type of measurement in which the assignment of numbers permit one to order objects or people along a continuum or dimension. In addition, numerically equal distances on the number scale correspond to equal amounts on the underlying dimension.

Ratio Measurement: Type of measurement that has all the properties of interval measures but which also map onto the underlying dimension in such a way that ratios between the numbers represent ratios of the dimension being measured.

Quantitative Variables: Variables measured on ordinal, interval, or ratio levels.

Qualitative Variables: Variables measured on a nominal level.

Scales: A term often used to refer to a measuring device or instrument used to obtain a set of measures.

Discrete Variables: Variables that can assume only a finite number of values or that have a finite number of values that can occur between any two points.

Continuous Variables: Variables that have an infinite number of values between any two points.

Real Limits: Points falling one-half a measure unit above a number (the upper real limit) and one-half a measure unit below that number (the lower real limit)..

Population: The aggregate of all cases to which one wishes to generalize.

Sample: A subset of the population.

Representative Sample: A sample that has the same basic characteristics as the population on a set of variables of interest.

Random Sampling: A strategy for selecting samples from a population. Every member of the population has an equal chance of being selected for the sample.

Random Number Table: A table of numbers that have been generated by a computer so that they closely approximate a set of random numbers.

Parameters: Numerical indices based on data from an entire population.

Statistics: Numerical indices based on data from a sample.

Descriptive Statistics: The use of numerical indices to describe either a population or a sample.

Inferential Statistics: The practice of taking measurements on a sample and then, from these observations, inferring the value of a population parameter.

Probability: The number of outcomes favoring an event divided by the total number of possible outcomes.

Summation Notation: A shorthand way of writing an instruction to sum a set of scores or perform other algebraic manipulations on the scores.

Practice Questions: True-False Format

1. Scientific research is best conceptualized as a body of knowledge about some phenomenon.

2. The scientific research process is always an orderly sequence of activities.

3. A hypothesis involves drawing a conclusion and thinking about the implications of the investigation for future research.

4. Most behavioral science research is concerned with constants.

5. A variable is a phenomenon that takes on different values or levels.

6. In a study on the effects of stress on alcohol consumption, stress is the independent variable.

7. The independent variable is the effect and the dependent variable is the cause.

8. It is necessary for a variable to be manipulated in order for it to be conceptualized as an independent variable.

9. Measurement involves translating empirical relationships between objects into numerical relationships.

10. Nominal measurement involves using numbers merely as labels.

11. Ordinal measures provide information about the magnitude of the differences between the objects.

12. Ordinal, interval, and ratio measures must occur on scales that are restricted to positive integers.

13. Most measures in behavioral science research are of nominal, ordinal, or interval level, with relatively few being ratio level.

14. Variables measured on a nominal level are called qualitative variables.

15. A measure has as its referent only a particular scale.

16. The determination of whether a variable is measured on an ordinal, interval, or ratio level is usually a straightforward matter in the behavioral sciences.

17. The number of people who attend an anxiety clinic is an example of a discrete variable.

18. Intelligence is an example of a continuous variable.

19. The real limits of a number are those points falling one measurement unit above and one measurement unit below that number.

20. A sample is an aggregate of all cases to which one wishes to generalize.

21. On the basis of observing the population, the researcher makes generalizations to the sample.

22. A biased sample is one that will lead us to make erroneous statements about the population.

23. Scientists use representative sampling to obtain random samples.

24. The essential characteristic of random sampling is that every member of the population has an equal chance of being selected for the sample.

25. Random sampling is an ideal that is seldom achieved in practice.

26. The use of random sampling procedures guarantees that a sample will be representative of the population.

27. Numerical indices based on data from a sample are referred to as statistics.

28. Inferential statistics involve the use of population parameters to estimate sample statistics.

29. A probability must always range from 0 to 1.00.

30. ΣX tells us to add up, or sum, all of the individual X scores.

31. 10.345 rounded to two decimals is 10.34.

Use the following scores for questions 32-35.

Individual	X	Y
1	3	5
2	2	2
3	1	3
4	4	1

32. $\Sigma X = 12$

33. $\Sigma X^2 = 30$

34. $\Sigma XY = 26$

35. $(\Sigma X)^2 = 900$

Answers to True-False Items

1. F	11. F	21. F	31. T
2. F	12. F	22. T	32. F
3. F	13. T	23. F	33. T
4. F	14. T	24. T	34. T
5. T	15. F	25. T	35. F
6. T	16. F	26. F	
7. F	17. T	27. T	
8. F	18. T	28. F	
9. T	19. F	29. T	
10. T	20. F	30. T	

Practice Questions: Short Answer

1. List the five general activities that characterize scientific research.

2. Distinguish between an independent variable and a dependent variable.

3. What is measurement?

4. Briefly describe the four types of measurement typically used in the behavioral sciences.

5. Distinguish between quantitative and qualitative variables.

6. Distinguish between discrete and continuous variables.

7. Distinguish between a population and a sample. Give an example of each.

8. Describe the relationship between random sampling and representative samples.

9. Distinguish between parameters and statistics.

10. Distinguish between descriptive and inferential statistics.

Answers to Short Answer Questions

1. Scientific research is characterized as an ongoing process consisting of five general activities: a) formulation of a question about some phenomenon or phenomena; b) forming a hypothesis concerning the question; c) designing an investigation to test the validity of the hypothesis; d) analyzing the data collected in the investigation; and e) drawing a conclusion and thinking about the implications of the investigation for future research.

2. The independent variable is assumed to influence the dependent variable. In cause and effect thinking, the independent variable is the cause and the dependent variable is the effect.

3. Measurement involves translating empirical relationships between objects into numerical relationships. This takes the form of assigning numbers to people (or objects) in such a way that the numbers have meaning and convey information about differences between people.

4. Nominal measurement involves using numbers merely as labels. Ordinal measurement involves the ordering of categories on some continuum or dimension. Interval measures have all the properties of ordinal measures, and in addition provide information about the magnitude of differences between the objects. Ratio measures have all the properties of interval measures, and in addition allow us to use ratios between the numbers to represent ratios of the dimension being measured.

5. Variables measured on ordinal, interval, or ratio levels are known as quantitative variables, while variables measured on a nominal level are called qualitative variables.

6. Variables that can assume only a finite number of values or that have a finite number of values that can occur between any two points are called discrete variables. A continuous variable can theoretically include an infinite number of values between any two points.

7. A population is the aggregate of all cases to which one wishes to generalize. For example, one might wish to study all married couples in the United States. A sample is simply a subset of the population. For example, one might randomly select 100 married couples to study. On the basis of observing the sample, the researcher makes generalizations to the population.

8. The essential characteristic of random sampling is that every member of the population has an equal chance of being selected for the sample. The use of random sampling is one technique used to approximate a representative sample.

9. Parameters are numerical indices based on data from an entire population. Statistics are numerical indices based on data from a sample.

10. Descriptive statistics involve the use of numerical indices to describe either a population or a sample. Inferential statistics involve taking measurements on a sample and then from the observations, inferring something about the population.

Chapter 2: Frequency and Probability Distributions

Study Objectives

This chapter describes methods for summarizing a set of scores. A primary tool for doing this is a frequency distribution. After reading this chapter, you should be able to define what a frequency distribution is and know how to construct a frequency distribution for a set of scores. You should know the difference between a frequency, a relative frequency, a cumulative frequency and a cumulative relative frequency and what information each of these convey. You should know how to construct grouped frequency distributions. You should be able to define an "outlier" and state why the identification of outliers is important.

The chapter also discusses methods for representing frequency distributions with graphs. You should be able to graph a frequency distribution using either a frequency histogram, frequency polygon, a line plot, a stem and leaf plot or a bar graph. You should also be able to graph frequency data for two or more groups.

You should understand the basic concept of a probability distribution, for both discrete variables and continuous variables. You should know the distinction between an empirical distribution and a theoretical distribution and be able to describe, in simple terms, a normal distribution.

Finally, you should be able to present a frequency distribution using the guidelines of the American Psychological Association (APA).

Study Tips

Statistics is very precise in its use of terms and vocabulary. It is important that you do not confuse terms, even though they sound very similar. For example, a common error that students make is to confuse the concepts of cumulative frequency and cumulative relative frequency. These terms refer to different constructs and you should be especially careful not to mix them up. Another common error that students make when constructing frequency graphs is to mistakenly put the frequencies on the abscissa and the scores of the independent variable on the ordinate. The reverse is typically done (except for some forms of the stem and leaf plot; see Figure 2.8a in the textbook).

Glossary Of Important Terms

Study the terms listed below. Make sure you understand each so that you could explain them to someone else who does not know them.

Frequency Distribution: A table that lists scores on a variable in order and shows the number of individuals who obtained each score or value.

Frequency: The number of times that a score value occurs in a set of scores

Relative Frequencies: The number of times that a score value occurs in a set of scores divided by the total number of scores.

Percentage: A relative frequency multiplied by 100.

Cumulative Frequencies: Frequency associated with a given score plus the sum of all frequencies for scores less than that score.

Cumulative Relative Frequencies: The relative frequency associated with a given score plus the sum of all relative frequencies for scores less than that score.

Cumulative Percentage: A cumulative relative frequency times 100.

Grouped Frequency Distributions: A table of a frequency distribution; however scores on the variable of interest are grouped together and each group is "treated as one."

Outliers: A case or set of cases that shows a very extreme score relative to the majority of cases in the data set - so extreme that the score is suspect.

Abscissa: The horizontal dimension of a graph; the X axis.

Ordinate: The vertical dimension of a graph; the Y axis.

Frequency Histogram: A graph of a frequency distribution for a quantitative variable in which frequencies occur on the ordinate and scores occur on the abscissa. The frequencies are denoted with bars that occur side by side.

Frequency Polygon: Similar to a frequency histogram, however bars are not used, but rather, solid dots corresponding to the appropriate frequencies are placed directly above the score values. The dots are connected with a solid line and the polygon is "closed" on the abscissa with boundary frequencies of zero.

Line Plot: Similar to a frequency polygon, except the line connecting the dots is not "closed" on the abscissa. Rather, the left and right most portions of the line end on the lowest and highest scores, respectively.

Stem and Leaf Plot: A type of frequency graph in which scores are listed on the ordinate and represented in such a way that every score in the distribution appears on the graph

Bar Graph: A frequency histogram for qualitative variables; bars are drawn such that they do not touch one another.

Probability Distribution: A distribution of probabilities across values of a variable

Probability Density Function: A method for representing the probability of observing scores between any two values for a continuous variable.

Density Curve: The area under the curve in a probability distribution representation of a continuous variable.

Empirical Distributions: Distributions of scores based on actual measurements collected in the real world.

Theoretical Distributions: Distributions of scores based on a mathematical function.

Normal Distribution: A symmetrical distribution that is characterized by a "bell shape."

Practice Questions: True-False Format

1. A frequency distribution is a table that lists scores on a variable and shows the number of individuals who obtained each score.

2. There are numerous hard-and-fast rules for presenting frequency information.

3. Considered alone, an index of frequency is always meaningful.

4. A relative frequency is the number of scores of a given value divided by the total number of scores.

5. When a relative frequency is multiplied by 100, it reflects the proportion of times the score occurred.

6. For any given score, the cumulative frequency is the frequency associated with that score plus the sum of all frequencies below that score.

7. The advantage of cumulative frequencies is that they allow us to tell at a glance the number of scores that are equal to or greater than a given score value.

8. Cumulative percentages indicate the percentage of cases that have scores equal to or less than a given score value.

9. Cumulative frequencies and cumulative relative frequencies are conceptualized with respect to the lower real limit of a score.

10. In constructing a frequency table, it would be neither practical nor informative to list 100 different values, each with a frequency of 1.

11. One should always report at least 10 groups in a grouped frequency distribution.

12. Typically, interval sizes of 2, 3, or multiples of 5 are used in grouped frequency distributions.

13. The conventional starting point for beginning the lowest interval in a grouped frequency distribution is 0.

14. The concepts of cumulative frequencies, cumulative relative frequencies, and cumulative percentages are not applicable to frequency distributions for qualitative variables.

15. An outlier is a case or set of cases that shows an average score relative to the majority of cases in the data set.

16. Outliers are always due to clerical errors.

17. In a frequency histogram, the ordinate represents the frequency with which each score occurred.

18. If a variable is continuous, the vertical boundaries of the bar for a given score will represent the absolute frequency of that score.

19. A frequency polygon is similar to a frequency histogram and uses the same ordinate and abscissa.

20. Frequency polygons differ from frequency histograms in that bars are not used, but rather, solid dots corresponding to the appropriate frequencies are placed directly above the score values.

21. Frequency polygons are typically used only when the variables being reported are discrete in nature.

22. Frequency histograms are typically used only when the variables being reported are continuous in nature.

23. A line plot is a type of stem and leaf plot

24. Frequency histograms and frequency polygons can be constructed for grouped as well as ungrouped scores.

25. A stem and leaf plot is used to graph probability distributions.

26. Stem and leaf plots are useful as long as the number of scores is not too large and when the number of different values on the base are moderate in number.

27. Because frequency graphs can be misleading depending on how the abscissa and ordinate are formatted, behavioral scientists rarely use them.

28. In a cumulative frequency graph, the cumulative frequency curve will always remain level or increase as it moves from left to right.

29. A probability represents the proportion of times that some score was previously observed.

30. When the potential values for a qualitative or discrete variable are mutually exclusive and exhaustive, then the probabilities associated with the individual score values will represent a probability distribution with respect to that variable.

31. Statisticians conceptualize a probability distribution of a discrete variable in terms of a probability density function.

32. The total area under a density curve represents 1.00.

33. Empirical and theoretical distributions refer to actual measurements collected in the real world.

34. All distributions in the family of normal distributions are symmetrical and are characterized by a "bell shape."

Answers to True-False Items

1. T	11. F	21. F	31. F
2. F	12. T	22. F	32. T
3. F	13. F	23. T	33. F
4. T	14. T	24. T	34. T
5. F	15. F	25. F	
6. T	16. F	26. T	
7. F	17. T	27. F	
8. T	18. F	28. T	
9. F	19. T	29. F	
10. T	20. T	30. T	

Practice Questions: Short Answer

1. Briefly describe what a frequency distribution is and how to construct one.

2. What is a relative frequency?

3. What is a cumulative frequency?

4. What three questions are central in deciding how to form the groups in a grouped frequency distribution?

5. How do we determine the number of groups to report in a grouped frequency distribution?

6. How do we determine the size of the interval to use in a grouped frequency distribution?

7. At what point do we begin the lowest interval in a grouped frequency distribution?

8. What is an outlier and why do they occur?

9. How does a frequency polygon differ from a frequency histogram?

10. What is the difference between a relative frequency and a probability?

11. Distinguish between empirical distributions and theoretical distributions.

12. What is the normal distribution?

Answers to Short Answer Questions

1. A frequency distribution is a table that lists scores on a variable and shows the number of individuals who obtained each value. We list the score values from highest to lowest. We then derive absolute frequencies by counting the number of individuals who received each score and indicate these frequencies next to the corresponding score values.

2. A relative frequency is the number of scores of a given value divided by the total number of scores--that is, the proportion of times that a score occurred.

3. Cumulative frequencies are obtained by a process of successive addition of the entries in the frequency column. For any given score, the cumulative frequency is the frequency associated with that score plus the sum of all frequencies below that score. The advantage of cumulative frequencies is that they allow us to tell at a glance the number of scores that are equal to or less than a given score value.

4. The three questions we must consider are: (1) How many groups should be reported? (2) What should the interval size be for each group? and (3) What should be the lowest value at which the first interval starts?

5. In deciding how many groups to report, a balance must be struck between having so many groups that the data are incomprehensible and having so few groups that the table is imprecise. In general, if the number of possible score values is small, fewer groups can be used, whereas if the number of possible score values is large, more groups will be required.

As a rule of thumb, the use of 5 to 15 groups tends to strike the appropriate balance between imprecision and incomprehensibility.

6. Typically, interval sizes of 2, 3, or multiples of 5 are used. To determine the interval size for a particular set of data, first subtract the lowest score from the highest score. This difference should then be divided by the desired number of groups and the result rounded to the nearest of the commonly used interval-size values.

7. The conventional starting point for the lowest interval in a grouped frequency distribution is the closest number evenly divisible by the interval size that is equal to or less than the lowest score.

8. An outlier is a case or set of cases that shows a very extreme score relative to the majority of cases in the data set--so extreme that the score is suspect. An outlier may be the result of a clerical error, as would be the case when a score is copied incorrectly from a questionnaire response. Alternatively, it may be that the person providing the response is somehow unique relative to the other people in the study.

9. A frequency polygon is similar to a frequency histogram and uses the same ordinate and abscissa. The major difference from the frequency histogram is that bars are not used, but rather, solid dots corresponding to the appropriate frequencies are placed directly above the score values. The dots are then connected by solid lines. Frequency polygons are always "closed" on the abscissa in the sense that they always include a value that is a unit higher than the highest observed score and a unit lower than the lowest observed score, with a zero frequency denoted for each.

10. Whereas a relative frequency indicates the proportion of times that some score was previously observed, a probability represents the likelihood of observing that score in the future.

11. Empirical distributions are based on actual measurements collected in the real world. Theoretical distributions are not constructed by formally taking measurements but, rather, are derived by making assumptions and representing these assumptions mathematically.

12. The normal distribution is one very important type of theoretical distribution that has been studied extensively by statisticians. There is actually a family of normal distributions, each member of which is precisely defined by a mathematical formula. All distributions in this family are symmetrical and are characterized by a "bell shape."

Chapter 3: Measures of Central Tendency and Variability

Study Objectives

This chapter presents methods for summarizing a set of scores. Based on the material in this chapter, you should be able to define and characterize the three major measures of central tendency, the mode, the median, and the mean. You should also be able to define and characterize the five major measures of variability, the range, the interquartile range, the sum of squares, the variance, and the standard deviation. You should be able to specify the strengths and weaknesses of each of these indices.

You should also be able to present graphs of central tendency and variability, using variants of bar graphs, line plots and boxplots.

You should be able to define skewness and kurtosis and recognize if a graph of a distribution is positively or negatively skewed. You show know what the terms leptokurtic and platykurtic mean.

You should know the difference between sample notation and population notation.

Finally, you should know how to present means and standard deviations using APA format.

Study Tips

Write down each of the formulas presented in this chapter on a separate piece of paper. Then, see if you can write out, *in words*, what the formula is telling you to do. It is important not to get too caught up in all the formulas, but rather to keep the "big picture" in mind. Students sometimes get so obsessed with executing formulas correctly that they forget about the concept that the formula is reflecting. The student emphasizes doing correct calculations, without thinking about what those calculations reflect and what the final product of the calculations mean. Do not fall into this trap. Always keep a conceptual focus. Ask yourself questions like: What does this number mean? Why would someone want to know this? Is the result a large or a small value? What are the implications of observing the result I calculated? Without an understanding of the *concepts* in this chapter, it will be difficult to comprehend later chapters. Make sure you have an intuitive feel for each concept.

This chapter presents symbols for concepts (e.g., s for standard deviation, s^2 for variance, SS for sum of squares, IQR for interquartile range, $\overline{X}$ for mean). Be sure that you can quickly and accurately associate a symbol with a construct.

Glossary Of Important Terms

Study the terms listed below. Make sure you understand each so that you could explain them to someone else who does not know them.

Central Tendency: An average of a set of scores; a value around which other scores tend to cluster.

Mode: The score that occurs most frequently.

Bimodal: When a distribution has two modes.

Median: The point in the distribution of scores that divides the distribution into two equal parts. Fifty percent of the scores occur above the median and 50% of the scores occur below the median.

Deviation Scores: The raw score minus the mean; it reflects how far a score deviates from the mean.

Mean: The sum of the scores divided by the sample size. It is the arithmetic average.

Unsigned Deviation Scores: The absolute value of the deviation score.

Variability: The extent to which scores are similar to one another.

Range: The highest score minus the lowest score.

Interquartile Range: The range of the scores after the top 25% of the scores and the bottom 25% of the scores have been "trimmed" or eliminated.

Sum of Squares: The sum of the squared deviation scores. It is an index of variability that, when equal to zero, means there is no variability. The higher the value of the sum of squares, the more variability there is, everything else being equal.

Variance: The sum of squares divided by the sample size. It is the average squared deviation from the mean (so it is sometimes called a **mean square**, for short). It is an index of variability that, when equal to zero, means there is no variability. The higher the value of the variance, the more variability there is, everything else being equal.

Standard Deviation: The positive square root of the variance. It is, roughly speaking, how far scores deviate, on average, from the mean. It is one of the more intuitive measures of variability. When it equals zero, there is no variability in the scores. The higher the value of the standard deviation, the more variability there is, everything else being equal.

Boxplot: A way of presenting measures of central tendency and variability graphically. It presents a box in the context of an X and Y axis, the height of which is related to the amount of variability.

Box and Whisker Plot: Another name for boxplot.

Skewness: The tendency for scores to cluster on one side of the mean.

Positively Skewed: The tendency for scores to cluster below the mean.

Negatively Skewed: The tendency for scores to cluster above the mean.

Kurtosis: The flatness or peakedness of one distribution relative to another.

Practice Questions: True-False Format

1. Central tendency refers to the "average" score in a set of scores.

2. The median of a distribution of scores is the score that occurs most frequently.

3. If we were to randomly select one score from a set of scores, the value of that score would most likely be equal to the mode as opposed to any other value.

4. The major problem with the mode as a measure of central tendency is that there can be only one modal score.

5. The median is the point in the distribution of scores that divides the distribution into two equal parts.

6. When there are duplications of the middle score(s), the median is simply the middle score.

7. The formula for computing the median is based on the assumption that the median occurs within the real limits of the middle score(s).

8. Across all individuals, scores will always tend to be closer to the mode than any other index of central tendency.

9. The mean of a set of scores is the arithmetic average of the scores.

10. The mean is computed by summing all of the scores and then dividing the sum by the squared deviation scores.

11. The sum of signed deviations about the mean will always equal 1.0.

12. The mean maximizes the sum of signed deviations.

13. In practice, the three measures of central tendency will almost always be equal to one another.

14. Most of the inferential statistics used by behavioral scientists make use of the median as a measure of central tendency.

15. Ideally, when trying to characterize a set of scores, it is useful to report all three indices of central tendency.

16. The more extreme a score is relative to the other scores in the distribution, the less it will change the value of the mean if it is deleted from the analysis.

17. When a quantitative variable is measured on an ordinal level that departs markedly from interval characteristics, the mean is the most meaningful measure of central tendency.

18. The concepts of median and mean are meaningless for qualitative variables.

19. The concepts of mean, median, and mode are applicable to continuous and discrete variables.

20. The variability of a set of scores reflects its skewness.

21. The range is the highest score minus the lowest score.

22. The range is a good measure of variability because it takes extreme scores into account.

23. The interquartile range is the difference between the highest and lowest scores after the top 75% of the scores and the bottom 75% of the scores have been eliminated from the data.

24. The sum of squares takes into account all of the scores in a data set.

25. The median is used as the measure of central tendency when defining the sum of squares.

26. When the mean and median are different, the sum of the squared deviations from the mean will always be less than the sum of the squared deviations from the median.

27. The sum of squared deviations from the mean will always be greater than the sum of squared deviations around any other index of central tendency.

28. One problem with the sum of squares as an index of variability is that its size depends not only on the amount of variability among scores, but also on the number of scores (N).

29. The variance is the average squared deviation from the mean.

30. The standard deviation is the positive square root of the sum of squares.

31. The standard deviation represents an average deviation from the mean.

32. When variables are measured on an ordinal level that departs markedly from interval characteristics, the standard deviation is the best measure of variability.

33. Skewness refers to the tendency for scores to cluster on one side of the mean.

34. Most scores in negatively skewed distributions occur below the mean and only a relatively few extreme scores occur above it.

35. Kurtosis refers to the flatness or peakedness of a distribution.

Answers to True-False Items

1. T	11. F	21. T	31. T
2. F	12. F	22. F	32. F
3. T	13. T	23. F	33. T
4. F	14. F	24. T	34. F
5. T	15. T	25. F	35. T
6. F	16. F	26. T	
7. T	17. F	27. F	
8. F	18. T	28. T	
9. T	19. T	29. T	
10. F	20. F	30. F	

Practice Questions: Short Answer

1. What is a central tendency?

2. Briefly define the mode and identify its weakness as a measure of central tendency.

3. Briefly define the median and describe the three different approaches to its computation.

4. Briefly define the mean and its relation to deviation scores.

5. Explain why the mean is not an optimal measure of central tendency when there are one or more extreme scores in the distribution.

6. What does the variability of a set of scores refer to?

7. What is the standard deviation?

8. What is the relationship between central tendency and variability?

9. Define skewness and describe what is meant by positively and negatively skewed distributions.

10. Define kurtosis and describe what is meant by platykurtic and leptokurtic distributions.

Answers to Short Answer Questions

1. A central tendency refers to an average, or a score around which other scores tend to cluster. Although there are many indices of central tendency, the three most commonly used in behavioral research are the mode, the median, and the mean.

2. The mode is the score that occurs most frequently. If the mode is used as the index of central tendency, then we are using the most common score as the "representative" value for the set of scores. The major problem with the mode is that there can be more than one mode.

3. The median is the point in the distribution of scores that divides the distribution into two equal parts. When there is an even number of scores, the median is the arithmetic average of the two middle scores. When there is an odd number of scores, the median is simply the middle score. When there are duplications of the middle score(s), a formula is used for computing the median based on interpolation within the real limits of the "middle" score.

4. The mean is the arithmetic average of a set of scores. It is computed by summing all of the scores and then dividing the sum by the total number of scores. The sum of signed deviations about the mean will always equal 0, that is, the mean is the value that minimizes the sum of signed deviations.

5. In general, when there are one or more extreme scores disproportionally on one side of the distribution, the information conveyed by the mean can become distorted and it loses its value as a score that "represents' the set of scores.

6. The variability of the scores reflects the extent to which they are similar to one another.

7. The standard deviation is the positive square root of the variance. It represents an average deviation from the mean.

8. Central tendency and variability represent different characteristics of a distribution. Distributions of scores can have identical variabilities but different central tendencies. Distributions can also have identical central tendencies but different variabilities. However, measures of variability can help to interpret measures of central tendency.

9. Skewness refers to the tendency for scores to cluster on one side of the mean. In positively skewed distributions, most scores occur below the mean and only a relatively few extreme scores occur above it. In negatively skewed distributions, most scores occur above the mean and only a relatively few extreme scores occur below it.

10. Kurtosis refers to the flatness or peakedness of one distribution relative to another. If a distribution is less peaked than another, it is said to be more platykurtic, and if it is more peaked than another, it is said to be more leptokurtic.

Answers to Selected Exercises from Textbook

Exercise 3: Because there are duplicate scores near the middle, we must use equation 3.1. We first write out the equation so as to identify what we need to compute:

$$\text{Median} = L + \left(\frac{(N)(.50) - n_L}{n_W} \right) i$$

We need to compute L, N, n_L, n_W, and i. First, we organize the data as a frequency distribution and then calculate the relative frequencies and cumulative relative frequencies:

Score	f	rf	crf
9	1	.20	1.00
6	1	.20	.80
4	3	.60	.60

Starting at the bottom of the crf column, we move up until we find the first number that is greater than or equal to .50. The number is .60, and occurs for the score of 4. Therefore,

L is the lower real limit of 4. This equals 3.5.

n_W is the number of scores (f) within the score category of 4. This equals 3.

N is the total number of scores. This equals 5.

n_L is the number of scores below L (i.e., 3.5). This equals 0.

i is the interval size of the score category 4, which is 3.5 to 4.5. Thus, I = 1.0.

Now we substitute in these values:

$$\text{Median} = 3.5 + \left(\frac{(5)(.50) - 0}{3} \right) 1$$

$$= 3.5 + \left(\frac{2.50}{3} \right) 1$$

$$= 3.5 + .83 = 4.33$$

The value that divides the distribution in half, taking into account the real limits of the scores, is 4.33.

Chapter 4: Percentiles, Percentile Ranks, Standard Scores, and the Normal Distribution

Study Objectives

This chapter presents methods for interpreting a single score in a distribution of scores. Based on the material in this chapter, you should be able to define and interpret a percentile and a percentile rank. You should be able to define and interpret a standard score. You should know the fundamental properties of standard scores (e.g., they have a mean of zero and a standard deviation of 1.0). You should know the basic characteristics of a normal distribution and be able to interpret a standard score within a normal distribution. For example, what proportion of cases in a normal distribution have a standard score greater than 1.96? Given a raw score, a mean, a standard deviation and knowledge that the scores are normally distributed, you should be able to calculate the proportion of scores that are greater than (or less than) that score.

Study Tips

Chapters 2 and 3 focused on statistical indices for describing an entire set of scores. By contrast, this chapter does not focus on describing a set of scores. Rather, it concerns how to interpret a *single* score within a distribution of scores. Students sometimes miss this change in focus and you should keep it in mind. The methods described in this chapter help you to determine how extreme a given person's score is. Do not confuse this with statistical indices to describe what a set of scores, considered as a whole, is like.

Glossary Of Important Terms

Study the terms listed below. Make sure you understand each so that you could explain them to someone else who does not know them.

Percentile Rank: The percentage of scores in a distribution that occur at or below a given value.

Relative Standing: The position of a score relative to other scores in the distribution.

Percentile: The score corresponding to a given percentile rank.

Standard Scores: A score that indicates the number of standard deviations that a raw score is above or below the mean.

Normal Distributions: A family of theoretical distributions that are defined by a mathematical formula. There is a different normal distribution for every combination of mean and variance. All normal distributions are symmetrical and bell shaped. The mean is equal to the median which is equal to the mode in a normal distribution.

z Score: A standard score in a normal distribution

T Score: A standard score that has been transformed so that it has a mean equal to an *a priori* specified value and a standard deviation equal to an *a priori* specified value. A T score avoids confusion associated with decimals and negative values.

Practice Questions: True-False Format

1. In general, observations are meaningful only in relation to other observations.

2. The percentage of scores in the distribution that occur at or below a given value, X, is the percentile of that value.

3. The score value corresponding to a given percentile rank is referred to as a standard score.

4. The score value corresponding to a percentile rank of 60 is referred to as the 60th percentile.

5. The median corresponds to the 50th percentile.

6. A percentile rank is an interval measure of relative standing.

7. To say that a score has a percentile rank of 80 is simply to state that 80% of the individuals scored at or above that score.

8. A standard score converts a score from its original, or raw, form to a form that takes into consideration its standing relative to the mean and standard deviation of the entire distribution of scores.

9. A standard score is the difference between the original score and the standard deviation, divided by the mean.

10. A standard score represents the number of standard deviation units that a score falls above or below the mean.

11. A positive standard score indicates that the original score is greater than the standard deviation.

12. A negative standard score indicates that the original score is less than the mean.

13. A standard score of 0 indicates that the original score is equal to the standard deviation.

14. The mean of a set of standard scores is always equal to 0.

15. The standard deviation of a set of standard scores is always equal to 1.00.

16. One important use of standard scores is to compare scores on distributions that have different means and standard deviations.

17. Not all normal distributions are symmetrical about the mean.

18. All normal distributions reflect empirically real phenomena.

19. All normal distributions are characterized by a "bell shape."

20. In most normal distributions, the mean, the median, and the mode are different from each other.

21. In some normal distributions, the proportion of scores that occur above or below a given standard score is different.

22. It is always the case that 50% of the scores in a normal distribution occur above the mean and 50% of the scores occur below the mean.

23. Knowing that a set of scores approximates a normal distribution allows us to make probability statements with respect to those scores, given that we know the mean and standard deviation.

24. The proportion of scores occurring between standard scores of 0 and -1 is less than the proportion of scores occurring between standard scores of 0 and +1.

25. In a normal distribution, 50% of all scores fall between standard scores of -1 and +1.

26. In a normal distribution, 20% of all scores fall between standard scores of -2 and +2.

27. In a normal distribution, over 99% of all scores fall between standard scores of -3 and +3.

28. A standard score in a normal distribution is referred to as a z score.

29. The process of standardizing scores changes the basic shape of the distribution.

30. A set of scores that is positively skewed will become normally distributed after they have been converted to standard scores.

31. A set of scores that is platykurtic will still be platykurtic after they have been converted to standard scores.

32. Standardization affects the central tendency, variability, skewness, and kurtosis of the scores.

33. T scores are directly analogous to standard scores, but instead of having a mean of 0 and a standard deviation of 1.00, they have a different mean and standard deviation and no negative values.

Answers to True-False Items

1. T	11. F	21. F	31. T
2. F	12. T	22. T	32. F
3. F	13. F	23. T	33. T
4. T	14. T	24. F	
5. T	15. T	25. F	
6. F	16. T	26. F	
7. F	17. F	27. T	
8. T	18. F	28. T	
9. F	19. T	29. F	
10. T	20. F	30. F	

Practice Questions: Short Answer

1. Distinguish between a percentile rank and a percentile.

2. What is a standard score?

3. What are some of the important properties of standard scores?

4. List four characteristics shared by all normal distributions.

5. Why are standard scores more meaningful when they occur in a normal distribution?

6. What is a z score?

7. How does the standardization of a set of scores affect the basic shape of the distribution?

8. What is a T score?

9. In what sense are percentile ranks and standard scores similar to each other?

10. What are some of the advantages of standard scores over percentile ranks?

Answers to Short Answer Questions

1. The percentage of scores in the distribution that occur at or below a given value, X, is the percentile rank of that value. The score value corresponding to a given percentile rank is referred to as a percentile.

2. A standard score represents the number of standard deviation units that a score falls above or below the mean. It summarizes the individual's relative standing, taking into consideration the mean and standard deviation of the distribution. It is the difference between the original score and the mean, divided by the standard deviation.

3. A positive standard score indicates that the original score is greater than the mean, and a negative standard score indicates that the original score is less than the mean. A standard score of 0 indicates that the original score is equal to the mean. The mean of a set of standard scores is always equal to 0. The standard deviation (and the variance) of a set of standard scores is always equal to 1.00.

4. All normal distributions are symmetrical about the mean, all are characterized by a "bell shape," and in all cases, the mean, the median, and the mode are equal. Another important feature of normal distributions is that they are theoretical in nature.

5. In normal distributions, the proportion of scores that occur above or below a given standard score is the same in all such distributions, as is the proportion of scores that occur between two specified standard scores. It is always the case, for example, that .50 of the scores in a normal distribution occur above the mean and .50 of the scores in a normal distribution occur below the mean. Thus, knowing that a set of scores approximates a normal distribution allows us to make probability statements with respect to those scores.

6. A z score is a standard score in a normal distribution.

7. The process of standardizing scores does not change the basic shape of the distribution. Standardization affects the central tendency and the variability of the scores, but not the skewness or kurtosis of the scores.

8. T scores are directly analogous to standard scores, but instead of having a mean of 0 and a standard deviation of 1.00, they have a mean of 50.00 and a standard deviation of 10.00.

9. Percentile ranks and standard scores are measures used to identify the location of a as specified score within a set of scores. They represent an individual's performance on some variable relative to some group of individuals. They do not convey performance in an absolute sense.

10. A percentile rank reflects only an ordinal measure of relative standing. A standard score takes into consideration a score's standing relative to the mean and standard deviation of the entire distribution of scores. Standard scores also enable us to compare scores on distributions that have different means and standard deviations.

Answers to Selected Exercises from Textbook

Exercise 4: The formula for calculating a percentile rank is

$$PR_X = \left(\frac{(.5)(n_W) + n_L}{N} \right) (100)$$

(a) For a score of 7, this yields

$$PR_7 = \frac{[(.5)(44) + 450]}{500} (100)$$

$$= \frac{472.00}{500} (100) = 94.40$$

(b) For a score of 5, we have

$$PR_5 = \frac{[(.5)(49) + 350]}{500} (100)$$

$$= \frac{374.50}{500} (100) = 74.90$$

(c) For a score of 3, we have

$$PR_3 = \frac{[(.5)(98) + 52]}{500} (100)$$

$$= \frac{101.00}{500} (100) = 20.20$$

(c) For a score of 2, we have

$$PR_3 = \frac{[(.5)\,(52) + 0]}{500}\,(100)$$

$$= \frac{26.00}{500}\,(100) = 5.20$$

Chapter 5: Pearson Correlation and Regression: Descriptive Aspects

Study Objectives

This chapter covers statistics for examining the relationship between two quantitative variables. It focuses on linear relationships. After reading this chapter, you should be able to define a slope and an intercept and interpret each. You should be able to define a correlation coefficient and interpret it in terms of its magnitude and how it characterizes the nature of the relationship between two variables (direct or inverse). You should know the maximum and minimum values of a correlation coefficient and what a correlation coefficient of zero means. You should know how to calculate a correlation coefficient from a set of data. You should know what a sum of cross products is and how it is involved in the computation of the correlation coefficient. You should know why two variables can be correlated even when there is no causal link between them.

You should know what a scatterplot is and be able to draw one for any set of X-Y scores.

You should be able to define a slope and an intercept in the context of data that is not perfectly linear and define the least squares criterion. You should know what a "predicted Y value" is. You should be able to define, interpret, and calculate the standard error of estimate.

You should know and be able to explain why the traditional correlation coefficient is not necessarily diagnostic of curvilinear relationships. You should know and be able to explain why the regression equation for predicting Y from X will not be the same as the regression equation for predicting X from Y. You should know how restricted range can affect the magnitude of the correlation coefficient. You should know how standardization of variables affects the slope and intercept in a regression equation. Finally, you should know how outliers can affect the magnitude and interpretation of correlation coefficients.

Study Tips

A common error that students make is to think that a negative correlation coefficient implies a weaker linear relationship than a positive correlation coefficient. For example, a correlation of -.75 is just a strong as a correlation of +.75. The only difference is that the former indicates an inverse relationship between two variables whereas the latter indicates a direct relationship between two variables. Be careful not to fall into this trap.

Another concept that students tend to have a difficult time with is the interpretation of the standard error of estimate. What is a large value and what is a small value? Unlike the correlation coefficient, this depends on the units of measurement. Suppose I tell you that when predicting the number of children that people want to have (Y) from education (X), I obtained a standard error of estimate of 4.0. The Y variable is in units of "children" and an average error of 4 "children" is substantial. However, suppose that I tell you that when predicting annual income (Y) from education (X), I obtain a standard error of estimate of 4.0. The Y variable is in units of "dollars" and an average error of $4 is trivial for predicting annual incomes. To interpret a standard error of estimate, you must know the "metric" (i.e., the units of measurement) of the Y variable. Keep this in mind.

At this point, the word "standard" appears in several of the terms we have used: standard deviation, standard score, standard normal distribution, and standard error of estimate. Each of these terms refers to something different. Be careful not to confuse them, as many students do.

Glossary Of Important Terms

Study the terms listed below. Make sure you understand each so that you could explain them to someone else who does not know them.

Pearson Correlation Coefficient: A statistical index that reflects the degree to which two variables approximate a linear relationship. Values can range from -1.00 to + 1.00. In general, the larger the absolute value of Pearson correlation, the closer the approximation to a linear relationship.

Scatterplot: A graph for visualizing the relationship between two quantitative variables. Scores for one of the variables (X) are listed on the abscissa and scores for the other variable (Y) are listed on the ordinate. A dot is placed in the graph where the X and Y scores intersect.

Slope: A parameter in the linear model. It indicates the number of units that the Y variable is predicted to change given a one unit change in X.

Positive Relationship: Relationship in which as scores on X increase, scores on Y also increase.

Direct Relationship: Same as a positive relationship.

Negative Relationship: Relationship in which as scores on X increase, scores on Y decrease.

Inverse Relationship: Same as a negative relationship.

Intercept: A parameter in the linear model. It is the predicted value of Y when X=0.

Linear Model: A mathematical formula that identifies parameters that define a linear relationship: $Y = a + bX$.

Sum of Cross-Products: A statistical index that is the sum of the product of the X scores times the Y scores.

Regression: A statistical technique that can be used to identify a "best fitting" line that describes the linear relationship between two quantitative variables

Least Squares Criterion: The criterion used to determine the values of the slope and intercept when trying to fit a linear model to a set of data. It minimizes the sum of the squared discrepancies between the predicted and observed Y scores.

Standard Error of Estimate: An index of predictive error that represents an average discrepancy between the predicted and observed Y scores.

Correlation Matrix: A set of correlation coefficients organized into rows and columns which shows correlations between all possible combinations of variables.

Practice Questions: True-False Format

1. The slope of a line indicates the number of units variable Y (the dependent variable) is predicted to change given a one unit change in X (the independent variable).

2. The value of a slope can be positive only.

3. In the case of a positive or direct relationship between X and Y, as scores on X increase, scores on Y increase.

4. The point at which a line intersects the Y axis when X = 1.0 is called the intercept.

5. Linear relationships can differ in terms of the values of their intercepts as well as in the values of their slopes.

6. The general form of the linear model is $Y = aX + bX$.

7. It is common in the behavioral sciences to observe a perfect linear relationship between two variables.

8. The correlation coefficient can range from -1.00 to 0 to +1.00.

9. The sign of the correlation coefficient indicates the magnitude of the linear approximation.

10. In the case of a perfect negative linear relationship, the standard scores on X and Y will be identical.

11. When there is no linear relationship between X and Y, the z scores on X will bear no consistent relationship to the z scores on Y, either in size or sign.

12. When the correlation between two variables is nonzero, the value of the sum of z score products will also be zero.

13. A sum of squares indicates the extent to which two sets of scores vary from one another, or covary.

14. The fact that two variables are correlated necessarily implies that one variable causes the other to vary as it does.

15. It is possible for two variables to be related to one another, but for no causal relationship to exist between them.

16. When two variables, X and Y, are correlated, three possible reasons for their correlation are that (1) X might cause Y, (2) Y might cause X, or (3) some additional variable(s) might cause both X and Y.

17. The correlation that has been found between the amount of violence watched on television and aggressive behavior in children proves that watching violent television programs causes aggressive behavior.

18. A correlation of .50 always represents a large correlation.

19. In the behavioral sciences where complex behaviors are studied, correlations in the .20 to .30 range (and the -.20 to -.30 range) are usually considered to be unimportant.

20. Correlations for the types of variables typically studied by behavioral scientists will seldom exceed -.40 or +.40 and will often be considerably smaller.

21. When two variables are not perfectly correlated, the statistical technique of regression can be used to identify a line that will fit the data points perfectly.

22. The formula for computing the intercept is $a = \overline{Y} - b\overline{X}$.

23. In a regression equation, the slope and intercept are defined so as to maximize the squared vertical distances that the data points, considered collectively, are from the regression line.

24. The criterion for deriving the values of the slope and intercept is formally known as the least squares criterion.

25. The standard error of estimate represents an average error across individuals in predicting scores on Y from the regression equation.

26. The absolute magnitude of the standard error of the estimate is not meaningful.

27. If X helps to predict Y, then the standard error of estimate will be smaller than the standard deviation of Y.

28. Pearson correlation will detect linear and nonlinear relationships.

29. The regression equation for predicting X from Y is the same as the regression equation for predicting Y from X.

30. From a statistical perspective, the designation of one variable as the X variable and one variable as the Y variable is arbitrary.

31. When a regression equation is computed using standardized X and Y scores, the slope will always equal 0.

32. Outliers are most likely to raise interpretational complexities when sample sizes are small.

33. Restricting the range of two variables will always lower the correlation between them.

Answers to True-False Items

1. T	11. T	21. F	31. F
2. F	12. T	22. T	32. T
3. T	13. F	23. F	33. F
4. F	14. F	24. T	
5. T	15. T	25. T	
6. F	16. T	26. F	
7. F	17. F	27. T	
8. T	18. F	28. F	
9. F	19. F	29. F	
10. F	20. T	30. T	

Practice Questions: Short Answer

1. What does the slope of a line refer to?

2. What is the Pearson correlation coefficient?

3. Explain why a correlation between two variables does not necessarily imply that one variable causes the other to vary as it does.

4. What is a predicted value of Y?

5. What is the least squares criterion?

6. Define the standard error of estimate and discuss how it is interpreted.

7. What are the consequences of using the correlation coefficient to assess a curvilinear relationship between two variables?

8. What are the statistical and conceptual implications of designating one variable as the X variable and one variable as the Y variable?

9. What is restricted range and how does it affect the correlation coefficient?

10. What are outliers and how do they effect the correlation coefficient?

Answers to Short Answer Questions

1. The slope of a line indicates the number of units variable Y is predicted to change given a one unit change in variable X.

2. The Pearson correlation coefficient, represented by the letter r, indexes the extent of linear approximation between two variables. The correlation coefficient can range from -1.00 through 0 to +1.00. The greater r is in either a positive or a negative direction from 0, the better is the approximation. The sign of the correlation coefficient indicates the direction of the linear approximation. A positive correlation coefficient indicates a positive or direct relationship between X and Y. A negative correlation coefficient indicates a negative or inverse relationship between X and Y.

3. It is possible for two variables to be related to one another, but for no causal relationship to exist between them. Three possibilities are that (1) X might cause Y, (2) Y might cause X, or (3) some additional variable(s) might cause both X and Y creating a correlation between them.

4. A predicted value of Y is the value of Y that an individual is predicted to have given the individual's X and substituting the value of X into the linear equation $Y = a + bX$.

5. The criterion for deriving the values of the slope and intercept is formally known as the least squares criterion. The least squares criterion defines the slope and the intercept so as to minimize the sum of the squared discrepancies between the predicted and observed Y scores.

6. The standard error of estimate represents an average error across individuals in predicting scores on Y from the regression equation. There are two perspectives in interpreting the standard error of estimate. First, its absolute magnitude is meaningful. Second, the standard error of estimate can be compared with the standard deviation of Y. The standard deviation of Y indicates what the average error in prediction would be if one were to predict for each individual a Y score equal to the mean of Y. If X helps to predict Y, then the standard error of estimate will be smaller than the standard deviation of Y. The better the predictor X is, the smaller the standard error of estimate will be, everything else being equal.

7. Two variables might be related, but if they are related in a fashion that is nonlinear, Pearson correlation will not necessarily be sensitive to this. This is because Pearson correlation assesses only linear relationships. Thus, a Pearson correlation coefficient of 0 might be obtained even when two variables show a strong curvilinear relationship.

8. From a statistical perspective, the designation of one variable as the X variable and one variable as the Y variable is arbitrary--it is as easy to derive the line for predicting the first variable from the second as it is to derive the line for predicting the second variable from the first. From a conceptual perspective, however, the tradition is to designate the independent variable as the X variable and the dependent variable as the Y variable. The choice of the labels therefore has important implications.

9. Restricted range involves examining only a portion of the range of a variable. Depending on the particular circumstances, the magnitude of the correlation when a limited portion of this range is considered might be either lesser or greater than if the range had not been so restricted.

10. An outlier is an "aberrant" case. Just as outliers can turn a weak correlation into a strong correlation, so can outliers turn a strong correlation into a weak one. Outliers are most likely to raise interpretational complexities when sample sizes are small.

Answers to Selected Exercises from Textbook

Exercise 13: Exercise 13 tells us to compute the Pearson correlation coefficient for a set of data. Note that in the exercise there are 10 individuals whose behavior is measured with two behavioral measures, X and Y, for each individual. First, write the formula for the Pearson correlation coefficient to see what we need to compute. Since we have raw data, we use the computational formula for the Pearson correlation coefficient (Equation 5.8).

$$r = \frac{\Sigma XY - \dfrac{(\Sigma X)(\Sigma Y)}{N}}{\sqrt{\left(\Sigma X^2 - \dfrac{(\Sigma X)^2}{N}\right)\left(\Sigma Y^2 - \dfrac{(\Sigma Y)^2}{N}\right)}}$$

We need to compute ΣX, ΣY, ΣX^2, ΣY^2, and ΣXY.

Individual	X	Y	X^2	Y^2	XY
1	3	7	9	49	21
2	8	9	64	81	72
3	3	3	9	9	9
4	2	8	4	64	16
5	6	8	36	64	48
6	6	9	36	81	54
7	8	6	64	36	48
8	5	4	25	16	20
9	7	2	49	4	14
10	2	4	4	16	8

$$\Sigma X = 50 \quad \Sigma Y = 60 \quad \Sigma X^2 = 300 \quad \Sigma Y^2 = 420 \quad \Sigma XY = 310$$

Now we can substitute the values into the formula.

$$r = \frac{310 - \dfrac{(50)(60)}{10}}{\sqrt{\left(300 - \dfrac{(50)^2}{10}\right)\left(420 - \dfrac{(60)^2}{10}\right)}}$$

$$= \frac{310 - 300}{\sqrt{(50)(60)}} = .18$$

A correlation coefficient of .18 indicates a weak positive linear relationship between the two variables.

Exercise 19. Exercise 19 tells us to compute the regression equation for the same set of data. We need several intermediate statistics in order to calculate the regression equation. First, write the formula for the regression equation to see what we need to compute. This is Equation 5.9:

$$\hat{Y} = a + bX$$

We need to compute b, the slope of the regression line, and a, the intercept. The slope, b, can be computed from Equation 5.10.

$$b \ = \ \frac{SCP}{SS_X}$$

To compute b, we need the sum of the cross-products, and the sum of squares for the X scores. The sum of the cross products is computed using equation 5.7:

$$SCP \ = \ \Sigma XY \ - \ \frac{(\Sigma X)(\Sigma Y)}{N}$$

$$= \ 310 - \ \frac{(50)\,(60)}{10} \ = \ 10$$

The sum of squares for X is our standard computational formula for a sum of squares:

$$SS_X = \ \Sigma X^2 \ - \ \frac{(\Sigma X)^2}{N}$$

$$SS_X = 300 - \frac{(50)^2}{10} = 50$$

Therefore, the slope of the regression line is

$$b = \frac{SCP}{SS_X}$$

$$= \frac{10}{50} = .20$$

The intercept, a, can be computed from equation 5.11:

$$a = \overline{Y} - b\overline{X}$$

To find a, we must compute the mean for the X scores and for the Y scores.

$$\overline{X} = \frac{\Sigma X}{N} \qquad \overline{Y} = \frac{\Sigma Y}{N}$$

$$\overline{X} = \frac{50}{10} \qquad \overline{Y} = \frac{60}{10}$$

Thus the intercept is:

$$a = 6.00 - (.20)(5.00)$$

$$= 6.00 - 1.00 = 5.00$$

Now we substitute the values of the intercept and slope to yield the regression equation:

$$\hat{Y} = a + bX$$

$$\hat{Y} = 5.00 + .20X$$

If we plotted the regression line on a graph, the line would intersect the Y axis at the value of the intercept, 5.00; the slope of the line would be such that when X increased by one unit, Y increased by .20 units.

Chapter 6: Probability

Study Objectives

This chapter covers the basics of probability theory. After reading the chapter, you should be able to define a probability of a simple event, a conditional probability and a joint probability and derive these from a contingency table. You should be familiar with relationships between different probabilities. You should be able to define independence. You should be able to explain sampling with replacement and sampling without replacement and how the two strategies affect the probability of observing an event. You should be able to explain and use counting rules using permutations and combinations. You should be able to use the binomial expression and know how to apply it to hypothesis testing.

Study Tips

One of the most common mistakes that students make in the material covered in this chapter is to confuse the concepts of simple probability, joint probability and conditional probability. Make a special effort to remember the distinctions between these concepts. Some students also have difficulty with the counting rules. Practice these using the exercises from the textbook.

Glossary Of Important Terms

Study the terms listed below. Make sure you understand each so that you could explain them to someone else who does not know them.

Probability: The number of outcomes favoring an event divided by the total number of possible outcomes.

Trial: The act of creating the event.

Event: Each unique outcome from a trial.

Contingency Table: A table of numbers in which the columns are the values of one variable and the rows are the values of a second variable. The entry in a given cell of the table is the frequency with which the combination of values defined by the column and row occurs.

Cell: Each unique combination of values of two variables in a contingency table.

Marginal Frequencies: The sums of the frequencies in the corresponding row or column of a contingency table.

Mutually Exclusive: A mutually exclusive category is one which a given individual can be classified in one and only one category. For example, if the category is gender, then a person can be classified as male or female, but not both.

Exhaustive: A set of outcomes that includes all the possible outcomes that could occur.

Probability Distribution: A distribution of probabilities across values of a variable

Conditional Probability: The probability of an event occurring given that some other event has occurred.

Independence: Events A is independent of event B if the occurrence of B does not affect the probability of the event A.

Joint Probability: The likelihood of observing each of two events.

Sampling With Replacement: Replacing cases back into the population before another case is randomly selected.

Sampling Without Replacement: Selecting a case at random and then, without replacing this case, selecting another case at random.

Permutation: An ordered sequence of a set of objects or events. AB and BA are two different permutations or orderings.

Combination: A set of objects or events in which the ordering of objects is irrelevant. AB is the same combination as BA.

Factorial: $n! = (n)(n-1)(n-2)(n-3)...(1)$

Binomial Expression: In a sequence of n independent trials, each of which has only two possible outcomes (arbitrarily called a "success" and a "failure"), with the probability of p of success and the probability q of failure, the probability of r successes in n trials is defined by the binomial expression (see equation 6.12 in the text).

Hypothesis Testing: The act of specifying an expected result of a study assuming the null hypothesis is true. If our observations are sufficiently discrepant from the expected result that the difference cannot be attributed to chance, then the null hypothesis is rejected. Otherwise, we fail to reject the null hypothesis.

Null Hypothesis: The hypothesis that is assumed to be true for purposes of conducting a statistical test. It is typically a hypothesis of no difference, no effect, or no relationship.

Alternative Hypothesis: A competing proposal to the null hypothesis. If the null hypothesis states that the number of events one should observe is 10, then the alternative hypothesis might state that the number of events one should observe is not 10.

Alpha Level: A probability value that is used to determine whether a result can be attributed to chance or non-chance in the context of hypothesis testing.

Correction for Continuity: A correction factor for applications of the binomial distribution for cases of small number of events to create a closer approximation to the normal distribution.

Practice Questions: True-False Format

1. The concept of probability forms the foundation of inferential statistics as well as several descriptive statistical methods.

2. In the language of probability theory, the act of flipping a coin is called an event and each unique outcome is called a trial.

3. If we randomly select an individual from a population consisting of 60 males and 40 females, the probability of selecting a male is 60/100 = .60.

4. Each unique combination of variables in a contingency table is referred to as a contingent frequency.

5. The sum of the frequencies in the corresponding row or column of a contingency table are referred to as marginal frequencies.

6. Two outcomes are said to be mutually exclusive when it is possible for both outcomes to occur for a given individual.

7. Given a set of outcomes that are mutually exclusive and exhaustive, the sum of the probabilities of the outcomes will always equal 1.00.

8. A set of mutually exclusive and exhaustive outcomes and their associated probabilities is known as a probability distribution.

9. A conditional probability indicates the likelihood that an event will occur given that some other event occurs.

10. An event A is said to be independent of an event B if $p(A) = p(B)$.

11. When two events are related, this means that one causes the other.

12. A joint probability is the likelihood of a causal relationship between two events.

13. A joint probability can be represented as $p(A, B)$, where $p(A, B)$ stands for the probability of both event A and event B occurring.

14. In general, the probability of observing at least one of event A and event B is $p(A \text{ or } B) = p(A) + p(B) - p(A, B)$.

15. There are no mathematical relationships among probabilities of simple events, conditional probabilities, and joint probabilities.

16. Sampling with replacement involves the random selection of a given case and the replacement of that case with a different case from the same population.

17. In the context of probability theory, there is no difference between sampling with replacement and sampling without replacement.

18. Sampling without replacement involves selecting a case at random and then, without replacing this case, selecting another case at random.

19. The method of sampling--with versus without replacement--never has an effect on the probability of observing some event.

20. If the size of the sample is small relative to the size of the population, sampling with versus without replacement will not affect probabilities appreciably.

21. A permutation of a set of objects or events is a sequence in which the internal ordering of elements is irrelevant.

22. A combination of a set of objects or events is an ordered sequence.

23. In factorial notation, the expression 0! equals 1.

24. In factorial notation, the expression 3! = (3)(2)(1) = 6.

25. Assuming the null hypothesis is false, we can easily specify an expected result.

26. If our observations are so discrepant from the expected result that the difference cannot be attributed to chance, we will reject the null hypothesis in favor of the alternative hypothesis.

27. If the observed result of an investigation is similar enough to the outcome stated in the null hypothesis such that it can reasonably be attributed to chance, we will reject the null hypothesis.

28. Determining what constitutes chance versus non-chance findings under the assumption that the null hypothesis is true is made with reference to a probability value known as the alpha level.

29. When the alpha level is .05, a result is defined as non-chance if the probability of obtaining that result, assuming the null hypothesis is true, is greater than .05.

30. The correspondence between the binomial and normal distributions improves as n increases and as p becomes closer to .50.

31. The mean of a binomial distribution is equal to μ.

Answers to True-False Items

1. T	11. F	21. F	31. F
2. F	12. F	22. F	
3. T	13. T	23. T	
4. F	14. T	24. T	
5. T	15. F	25. F	
6. F	16. F	26. T	
7. T	17. F	27. F	
8. T	18. T	28. T	
9. T	19. F	29. F	
10. F	20. T	30. T	

Practice Questions: Short Answer

1. Using the example of a coin flip, distinguish between a trial and an event.

2. What does it mean to say that two outcomes are mutually exclusive?

3. What is a conditional probability?

4. What does it mean to say that two events are independent?

5. What is a joint probability?

6. Distinguish between sampling with replacement versus sampling without replacement.

7. What is the difference between a permutation and a combination?

8. Using the example of an experiment on extrasensory perception (ESP) where an individual's claim that she possesses psychic powers is tested by asking her to predict the outcome of each of 10 tosses of a coin, distinguish between a null hypothesis and an alternative hypothesis.

9. What is an alpha level and how does it relate to hypothesis testing?

10. What is the relationship between the binomial distribution and the normal distribution?

Answers to Short Answer Questions

1. In probability theory, the act of flipping the coin is called a trial and each unique outcome is called an event.

2. Two outcomes are said to be mutually exclusive when it is impossible for both outcomes to occur for a given individual. For example, the variable of gender has two possible outcomes: (1) being a male or (2) being a female. If a person is classified as being a male, he cannot also be classified as being a female. Thus, these outcomes are mutually exclusive.

3. A conditional probability indicates the likelihood that an event will occur given that some other event occurs. The general symbolic form for a conditional probability is $p(A|B)$, where $p(A|B)$ is read "the probability of event A, given event B."

4. An event A is said to be independent of an event B if $p(A) = p(A|B)$, that is, if the occurrence of event A is unrelated to the occurrence of event B. Two events are related (nonindependent) when the occurrence of event A is related to the occurrence of event B. It is important to realize that even though two events are related (nonindependent), this does not necessarily mean that one causes the other.

5. A joint probability refers to the likelihood of observing each of two events. A joint probability can be represented as $p(A, B)$, where $p(A, B)$ stands for the probability of both event A and event B occurring.

6. The process of sampling with replacement involves the random selection of a given case from a population, taking some measurement of interest, and returning that case to the population. Then, with the population fully intact, the random selection procedure can be performed again. In contrast, sampling without replacement involves selecting a case at random and then, without replacing this case, selecting another case at random, and so on. The method of sampling--with versus without replacement--can affect the probability of observing some event. When the size of the sample relative to the size of the population is large, the different sampling procedures can produce very different results. However, if the size of the sample is small relative to the size of the population, sampling with versus without replacement will not affect probabilities appreciable.

7. A permutation of a set of objects or events is an ordered sequence, whereas a combination of a set of objects or events is a sequence in which the internal ordering of elements is irrelevant. The general notation for the number of permutations (P) of n things taken r at a time is $_nP_r$. The general notation for the number of combinations (C) of n things taken r at a time is $_nC_r$.

8. In the context of hypothesis testing, the proposal, or hypothesis, that an individual does not possess psychic ability is called a null hypothesis. Assuming the null hypothesis is true, we can specify an expected result of an investigation. In our example, this takes the form that the individual will accurately predict 5 of the 10 coin tosses. If our observations are so discrepant from the expected result that the difference cannot be attributed to chance, we will reject the null hypothesis in lieu of a competing proposal, referred to as an alternative hypothesis. In our example, the alternative hypothesis is that the individual possesses the claimed psychic ability. On the other hand, if the observed result is similar enough to the outcome stated in the null hypothesis such that it can reasonably be attribute to chance, we will fail to reject the null hypothesis.

9. Within the context of hypothesis testing, the problem is one of determining what constitutes chance versus non-chance results under the assumption that the null hypothesis is true. This distinction is made with reference to a probability value known as an alpha level. For example, when the alpha level is .05, a result is defined as non-chance (that is, reflective of factors other than chance) if the probability of obtaining the result, assuming the null hypothesis is true, is less than .05.

10. When certain conditions are met, the normal distribution can be used to obtain a very close approximation to relevant binomial probabilities. It turns out that the binomial and normal distributions are closely related with the correspondence between them depending on the values of n and p. The correspondence improves as n increases and as p becomes closer to .50. For small n, statisticians have developed a correction factor (called the correction for continuity) that, when applied to the data, yields even better correspondence.

Answers to Selected Exercises from Textbook

Exercise 36: Twenty percent of individuals who seek psychotherapy will exhibit a return to normal personality irrespective of whether they receive treatment (spontaneous recovery). Thus, the probability of successes is p = .20 (i.e., 20% multiplied by .001), and the probability of failure is q = (1 - p) = (1 - .20) = .80. A researcher finds that a particular type of psychotherapy is successful with 30 out of 100 clients. Thus, n = 100, and we are seeking the binomial probability of a score of 30. The binomial probability of a score of 30 can be approximated by converting the score of 30 into a z score and finding the corresponding normal probability in Appendix B. First, write the z score formula to see what we need to compute (Equation 4.6).

$$z = \frac{X - \mu}{\sigma}$$

We need the mean of the binomial (μ), which is found by multiplying the number of trials (n) times the probability of success (p), and the standard deviation of the binomial (σ), which is the square root of the number of trials (n) times the probability of success (p) times the probability of failure (q). Therefore, we need to find n, p, and q:

$$n = 100$$
$$p = .20$$
$$q = .80$$

Now, substitute the numbers into the formula:

$$\mu = np$$

$$= (100)(.20) = 20.00$$

$$\sigma = \sqrt{npq} = \sqrt{(100)(.20)(.80)} = 4.00$$

We are now ready to solve for z:

$$z = \frac{30 - 20.00}{4.00} = 2.50$$

A score of 30 translates into a z score of 2.50. From Appendix B, the probability of obtaining a z score of 2.50 or greater is .0062. Since .0062 is less than the criterion value of .05, the researcher should conclude that the psychotherapeutic approach is more effective than no treatment in helping clients return to normal personality.

Chapter 7: Estimation and Sampling Distributions

Study Objectives

This chapter covers fundamental issues in estimating the value of a population parameter from sample data. After reading the material, you should be able to define sampling error, sampling distributions, and the standard error of the mean. You should be able to explain the concept of a sampling distribution and why it is important for inferential statistics. You should be able to define an unbiased estimator. You should understand the concept of degrees of freedom and its role in estimation. You should know the central limit theorem and be able to explain its implications for statistical inference. Finally, you should be able to discuss the major factors that influence sampling error.

Study Tips

This chapter is one of the most important in the book. It lays the foundation for all of the inferential statistics discussed in later chapters. It is important that you have a complete understanding of the material in this chapter, or else you will be lost in future chapters. Read it very carefully, make sure you master the material, and if *anything* is not clear, be sure to ask your instructor about it.

One common error that students make in this chapter is to assume that an unbiased estimator is one that has no sampling error. This is not true. Just because an estimator is unbiased does not mean it will accurately reflect the population parameter in any given sample. The term "unbiased" is a technical, statistical term and refers to a specific statistical property: the mean of the estimator across all possible random samples of a given size will equal the value of the population parameter.

The notion of a sampling distribution is an abstract concept and some students have difficulty with it. We never actually calculate sampling distributions in practice. However, we rely on the *concept* of a sampling distribution to help us derive other concepts (the standard error of the mean) and statistical estimates of those concepts (the estimated standard error of the mean) that are meaningful and which can be calculated in practice. Make sure you understand the logic of sampling distributions. It is fundamental.

At this point, it is very important that you keep in mind the distinction between populations, samples, and sample estimates of population parameters. The standard deviation in a population is not the same as a standard deviation in a sample, because the former is calculated for an entire population but the latter is calculated for only a sample from the population. A standard deviation estimate is not the same as a sample standard deviation. The former is a statistic that is used to *estimate* a population standard deviation from sample data. The latter is just a descriptor of what the standard deviation is in the sample data and does not represent a "guess" about what the population standard deviation is. The table on page 186 of your text is an extremely important one. If you do not understand it, be sure to ask your instructor.

Glossary Of Important Terms

Study the terms listed below. Make sure you understand each so that you could explain them to someone else who does not know them.

Population Parameter: A statistical index that describes a population based on all data from that population.

Sample Statistic: A statistical index based on data from a sample that can be used to infer something about a population or that can be used to describe a sample.

Sampling Error: Random or systematic factors that yield a sample statistic that does not equal the value of its corresponding population parameter.

Unbiased Estimator: A sample statistic whose average (mean) over all possible random samples of a given size equals the value of the corresponding population parameter.

Biased Estimator: A sample statistic whose average (mean) over all possible random samples of a given size does not equal the value of the corresponding population parameter. If the average is larger than the value of the population parameter, the estimator is said to be positively biased. If the average is smaller than the value of the population parameter, the estimator is said to be negatively biased.

Variance Estimate: An unbiased estimator of the population variance.

Standard Deviation Estimate: An unbiased estimator of the population standard deviation.

Degrees of Freedom: Used to indicate the number of pieces of information that are "free of each other" in the sense that they cannot be deduced from one another.

Mean Square: Any sum of squares divided by its associated degrees of freedom.

Sampling Distribution of the Mean: A distribution of mean scores, consisting of the means for all possible random samples of a given size from a population..

Central Limit Theorem: A theorem that describes characteristics of a sampling distribution of the mean: Given a population with a mean of μ and a standard of σ, then the sampling distribution has a mean of μ and a standard deviation of $\sigma/\sqrt{N}$ and approaches a normal distribution, as the sample size approaches infinity.

Standard Error of the Mean: The standard deviation of a sampling distribution of the mean.

Practice Questions: True-False Format

1. Estimating population parameters is relevant for small, finite populations as well as populations that are so large that for all practical purposes they can be considered infinite.

2. The statistics used in most behavioral science disciplines are not applicable to extremely large populations.

3. It is virtually impossible for investigators to select truly random samples from very large populations.

4. The term "sampling error" implies that mistakes have been made in the collection and analysis of the data.

5. The term "sampling error" reflects the fact that sample values are likely to differ from population values because they are based on only a portion of the overall population.

6. The amount of sampling error is almost always small.

7. In practice, it is possible to compute the exact amount of sampling error that occurs.

8. In the absence of any other information, the sample mean that one observes is one's "best estimate" of the value of the population mean.

9. If a statistic is an unbiased estimator, then it will always equal the value of the population parameter for any given sample.

10. In statistical terms, the sample mean is said to be a biased estimator of the population mean.

11. An unbiased estimator of a population parameter is one whose average (mean) over all possible random samples of a given size equals the value of the parameter.

12. Statisticians have determined that the sample variance is an unbiased estimator of the population variance.

13. The sample variance overestimates (is larger than) the population variance across all possible samples of a given size.

14. An unbiased estimator of the population variance cannot be obtained from sample data.

15. An unbiased estimate of the population variance can be obtained by dividing the sample sum of squares by N - 1.

16. By far the most common occurrence in the behavioral sciences involves the estimation of population parameters from sample data.

17. In statistics, the phrase degrees of freedom is used to indicate the number of pieces of information that are "free of each other" in the sense that they cannot be deduced from one another.

18. A sum of squares around a sample mean will always have N - 2 degrees of freedom associated with it.

19. In general, as the degrees of freedom associated with an estimate increase, the accuracy of the estimate also tends to increase.

20. Technically, the accuracy of a variance estimate is not a function of the degrees of freedom (N - 1), but rather is a function of the sample size (N) and the mean.

21. Irrespective of its specific computation, any sum of squares divided by its associated degrees of freedom is referred to as a mean standard deviation.

22. A sampling distribution of the mean is a theoretical distribution consisting of the mean scores for all possible random samples of a given size that could be drawn from a population.

23. In practice, we can always calculate a sampling distribution of the mean.

24. One result of the central limit theorem is that the mean of a sampling distribution of the mean is never equal to the population mean.

25. The standard deviation of a sampling distribution of the mean is called the standard error of the mean.

26. The standard error of the mean reflects the accuracy with which sample means estimate a population mean.

27. If the standard error of the mean is small, then all the sample means based on a given sample size (N) will be equal to the population mean.

28. The size of the standard error of the mean is influenced by two factors: the sample size and the variability of scores in the population.

29. A third implication of the central limit theorem is that the sampling distribution of the mean can be approximated by a binomial distribution when the sample size is sufficiently large.

30. When the degrees of freedom are greater than 40, the normal approximation of the sampling distribution of the mean is quite good.

31. There is a different sampling distribution of the mean for every sample size.

32. The standard error of the mean gets smaller as the sample size decreases.

33. Given the same population, the sampling distribution of the mean will show less variability than either the sampling distribution of the median or the sampling distribution of the mode.

Answers to True-False Items

1. T	11. T	21. F	31. T
2. F	12. F	22. T	32. F
3. T	13. F	23. F	33. T
4. F	14. F	24. F	
5. T	15. T	25. T	
6. F	16. T	26. T	
7. F	17. T	27. F	
8. T	18. F	28. T	
9. F	19. T	29. F	
10. F	20. F	30. T	

Practice Questions: Short Answer

1. In what sense are behavioral scientists concerned with infinite sized populations?

2. What is sampling error?

3. What is an unbiased estimator?

4. In what way is the sample variance a biased estimator of the population variance and how do we correct for this bias?

5. Explain what the phrase "degrees of freedom" refers to and discuss how it relates to the accuracy of an estimate.

6. What is a sampling distribution of the mean?

7. What is the central limit theorem?

8. Define the standard error of the mean.

9. What two factors influence the size of the standard error of the mean?

10. According to the central limit theorem, what is the relationship between the sampling distribution of the mean and the normal distribution?

Answers to Short Answer Questions

1. Behavioral science research is typically conducted with the goal of explaining the behavior of large numbers of individuals, often including people who have lived previously or who have yet to be born, as well as those residing in the present. As such, the statistics used in most behavioral science disciplines are applicable to extremely large, if not infinite, populations.

2. The fact that a sample statistic may not equal the value of its corresponding population parameter is said to be the result of sampling error. The term "sampling error" reflects the fact that sample values are likely to differ from population values because they are based on only a portion of the overall population. The amount of sampling error is the difference between the value of a sample statistic and the value of the corresponding population parameter.

3. An unbiased estimator of a population parameter is one whose average (mean) over all possible random samples of a given size equals the value of the population parameter. In statistical terms, the sample mean is said to be an unbiased estimator of the population mean.

4. Statisticians have determined that the sample variance is a biased estimator of the population variance in that it underestimates (is smaller than) the population variance across all possible samples of a given size. An unbiased estimator of the population variance, referred to as a variance estimate, can be obtained from sample data by dividing the sum of squares by N - 1 instead of N.

5. In statistics, the phrase "degrees of freedom" is used to indicate the number of pieces of information that are "free of each other" in the sense that they cannot be deduced from one another. In general, as the degrees of freedom associated with an estimate increase, the accuracy of the estimate also tends to increase.

6. A sampling distribution of the mean is a theoretical distribution consisting of mean scores for all possible random samples of a given size that could be drawn from a population.

7. Perspectives on the mean and standard deviation of a sampling distribution of the mean, as well as its shape, are derived from the central limit theorem. This theorem states: Given a population with a mean of μ and a standard deviation of σ, then the sampling distribution has a mean of μ and a standard deviation of $\sigma/\sqrt{N}$ and approaches a normal distribution, as the sample size approaches infinity.

8. The standard deviation of a sampling distribution of the mean is called the standard error of the mean. The standard error of the mean reflects the accuracy with which sample means estimate a population mean.

9. Two factors influence the size of the standard error of the mean. The first is the sample size. As the sample size increases, the standard error becomes smaller. The second factor is the variability of scores in the population. As σ becomes smaller, so does the standard error.

10. According to the central limit theorem, the sampling distribution of the mean more closely approximates a normal distribution as sample size increases distribution.

Answers to Selected Exercises from Textbook

Exercise 6: Exercise 6 first tells us to compute the variance and the standard deviation for a set of data. To compute the variance, we first need to compute the sum of squares using the standard formula (Equation 3.6).

$$SS = \Sigma X^2 - \frac{(\Sigma X)^2}{N}$$

We need to compute ΣX, N, and ΣX^2:

Score	X	X^2
1	2	4
2	3	9
3	3	9
4	4	16
5	4	16
6	4	16
7	5	25
8	5	25
9	5	25
10	5	25
11	6	36

Score	X	X²
12	6	36
13	6	36
14	7	49
15	7	49
16	8	64

$$\Sigma X = 80 \qquad \Sigma X^2 = 440$$

Now we can substitute these values into Equation 3.6 to obtain the sum of squares, Equation 3.4 to obtain the sample variance, and Equation 3.5 to obtain the sample standard deviation.

$$SS = \Sigma X^2 - \frac{(\Sigma X)^2}{N}$$

$$= 440 - \frac{(80)^2}{16} = 40.00$$

$$s^2 = \frac{SS}{N} = \frac{40.00}{16} = 2.50$$

$$s = \sqrt{s^2} = \sqrt{2.50} = 1.58$$

Because the sample variance is a biased estimator of the population variance, we apply Equation 7.1 to obtain an unbiased estimate of the population variance.

$$\hat{s}^2 = \frac{SS}{N-1}$$

$$= \frac{40.00}{16 - 1} = 2.67$$

To obtain an unbiased estimate of the population standard deviation, we apply Equation 7.2.

$$\hat{s} = \sqrt{\hat{s}^2}$$

$$= \sqrt{2.67} = 1.63$$

Notice that the estimates of the population variance and standard deviation are larger than the sample variance and standard deviation. Recall that the sample variance is a biased estimator of the population variance in that it underestimates the population variance across all possible samples of a given size. Thus, the denominator of Equation 3.4 is modified by dividing by N-1 instead of N. This "correction" involving the subtraction of 1 from N serves to make the variance estimate larger than the sample variance and, hence, corrects for the tendency of the sample variance to underestimate the population variance.

Exercise 21: This exercise tells us to compute the mean and the estimated standard error of the mean for the above data. We use the standard formula for the mean (Equation 3.2). Since we have sample data, we use the formula for the estimate of the standard error of the mean (Equation 7.5).

$$\overline{X} = \frac{\Sigma X}{N}$$

$$= \frac{80}{16} = 5.00$$

The mean of the distribution is 5.00. To compute the estimated standard error of the mean, we use Equation 7.5.

71

$$\hat{s}_{\overline{X}} = \frac{\hat{s}}{\sqrt{N}}$$

$$= \frac{1.63}{\sqrt{16}} = .41$$

Thus, the estimated standard error of the mean is .41. This means that, on average, for samples of size 16, sample means are estimated to differ .41 units from the true population mean.

Chapter 8: Hypothesis Testing: Inferences About a Single Mean

Study Objectives

This chapter introduces the basic logic of hypothesis testing. After reading the material, you should be able to summarize the steps involved in testing whether an observed mean is different from a population mean (1) under the scenario where the population variance is known, and (2) under the scenario where the population variance is not known. You should be able to define a null and alterative hypothesis and the concept of critical values and rejection regions. You should be able to define what an alpha level is and explain its role in hypothesis testing. You should be able to define Type I and Type II errors as well as the concept of statistical power. You should be able to explicate factors that affect statistical power and have an intuitive feel for *why* they affect statistical power.

You should be able to explain why we can never accept the null hypothesis. You should be able to distinguish between directional and non-directional tests and state the conditions where you would use each. You should know when to use the one sample z test as opposed to the one sample t test. You should be able to characterize the robustness of the one sample t test relative to its underlying assumptions. Finally, you should be able to explain the concept of confidence intervals and the logic underlying their computation. You should also know how to write-up a one sample t test using APA format.

Study Tips

Like Chapter 7, this chapter is one of the most important in the book. You must understand the logic of this chapter to understand the logic in later chapters. Some of the concepts are abstract and require careful thought. If something is not clear, be sure to ask your instructor for clarification.

This chapter builds extensively on the material in Chapters 4 and 7. Make sure that you are fluent with the use of terms and materials from these chapters before undertaking this one.

In some respects, a substantial portion of this chapter is learning a new vocabulary and keeping your definitions straight. Many new terms are introduced and you must keep distinctions between them clear. There is a tendency for students to focus on successfully performing the calculations required by the one sample t test (or z test), and losing sight of the bigger picture and what is going on conceptually. The result is that the student can calculate all of the right answers, but really does not know why he or she is doing the calculations and to what end. The concepts and logic from this chapter will be applied over and over again in later chapters, so it is important not to emphasize calculations over concepts.

Glossary Of Important Terms

Study the terms listed below. Make sure you understand each so that you could explain them to someone else who does not know them.

Hypothesis Testing: The act of specifying an expected result of a study assuming the null hypothesis is true. If our observations are sufficiently discrepant from the expected result that the difference cannot be attributed to chance, then the null hypothesis is rejected. Otherwise, we fail to reject the null hypothesis.

Null Hypothesis: The hypothesis that is assumed to be true for purposes of conducting a statistical test. It is typically a hypothesis of no difference, no effect, or no relationship. The symbol for the null hypothesis is H_0.

Alternative Hypothesis: A competing proposal to the null hypothesis. If the null hypothesis states that the number of events one should observe is 10, then the alternative hypothesis might state that the number of events one should observe is not 10.

One Sample z Test: A statistical test used to test the viability of the null hypothesis that the population mean is equal to a specific value. It is applied in cases where the population variance is known, which is rarely the case.

Critical Values: The critical values are the range of values within which the sample mean is expected to fall if sampling error is a reasonable explanation as to why the sample data are discrepant from the value of the mean specified in the null hypothesis. Critical values are usually specified in units of standard scores.

Rejection Region: Values of a test statistic that are outside the range of critical values. If the critical values are -1.96 to 1.96, then the rejection region refers to all values greater than 1.96 and less than -1.96.

Test Statistic: A statistical index calculated from sample data that is used to formally conduct a null hypothesis test (i.e., to evaluate the reasonableness of a sampling error interpretation of sample discrepancies from the null hypothesis). In this chapter, the z score and the t score are examples of test statistics.

Alpha Level: A probability value that is used to determine whether a result can be attributed to chance or non-chance in the context of hypothesis testing. The alpha level also reflects the probability of making a Type I error.

Type I Error: Rejecting the null hypothesis when it is true.

Type II Error: Failing to reject the null hypothesis when it is false.

Beta: The probability of making a Type II error.

Power: The probability of not making a Type II error. It equals one minus Beta. It is the probability that an investigator will correctly reject the null hypothesis when it is false.

Statistically Significant: A phrase that means that the null hypothesis was rejected based on the evaluation of a test statistic.

Statistically Nonsignificant: A phrase that means that the null hypothesis was not rejected based on the evaluation of a test statistic.

Directional Test: Statistical tests that are designed to detect differences from a hypothesized population value (as stated in the null hypothesis) in only one direction (e.g., the population mean is greater than the value specified in the null hypothesis).

One-Tailed Test: Same as a directional test.

Nondirectional Test: Statistical tests that are designed to detect differences from a hypothesized population value (as stated in the null hypothesis) in both directions (i.e., the population mean is greater than the value specified in the null hypothesis or it is less than the value specified in the null hypothesis).

Two-Tailed Tests: Same as a nondirectional test.

t Distribution: A theoretical distribution that is bell shaped and symmetrical, much like the normal distribution. Statisticians are able to make probability statements about the occurrence of different ranges of values in a t distribution, making it useful for statistical inference.

One-Sample t Test: A statistical procedure that involves the calculation of a test statistic (a t score) that has a sampling distribution that closely approximates a t distribution. The test statistic is used in conjunction with our knowledge of the t distribution to reject or not reject the null hypothesis.

Normality Assumption: The assumption that scores on the variable are normally distributed in the population.

Robust: A term that indicates that indicates how sensitive a statistical test is to violations of its underlying assumptions. A statistical test that is "robust" is said to not be affected (in terms of the incidence of Type I and Type II errors) by violations of its underlying assumptions.

Confidence Interval: A range of values within which we are relatively confident that the population parameter occurs. The confidence interval is stated in terms of a percentage (e.g., 95%) and indicates the percentage of times that the population mean falls within the confidence interval as computed in the sample data for all possible random samples of a given size.

Confidence Limits: The values that define the range of the confidence interval. The upper limit is the larger value and the lower limit is the smaller value.

Interval Estimation: The strategy of calculating and interpreting confidence intervals.

Probability Value (p Value): The probability of obtaining a result as extreme or more extreme than the one observed, given that the null hypothesis is true.

Practice Questions: True-False Format

1. When the variable under study is quantitative in nature and measured on a level that at least approximates interval characteristics, hypotheses about the value of a population mean can be tested using the one sample z test, if the population variance is known.

2. The null hypothesis is the hypothesis that we assume to be false for purposes of conducting a statistical test.

3. The set of all standard scores more extreme than the critical values is called a rejection region and constitutes an unexpected result.

4. One step in the hypothesis testing process is the conversion of the observed sample mean into a z value to determine how many standard errors it is away from μ, assuming the alternative hypothesis is true.

5. A test statistic is the same as an ordinary descriptive statistic and is not used inferentially.

6. Rejection regions are determined with reference to an alpha level.

7. For an alpha level of .05, an unexpected result in a one sample z test includes any sample mean occurring more than 1.96 standard errors below or 1.96 standard errors above the value of μ represented in the null hypothesis.

8. When a researcher obtains a result that is consistent with the null hypothesis, he or she accepts the null hypothesis as being true.

9. In principle, we can never accept the null hypothesis as being true via our statistical methods; we can only reject it as being untenable.

10. When the observed value of z falls within the range defined by the critical values, we reject the null hypothesis.

11. Rejection of the null hypothesis when it is true is called a Type II error.

12. Failure to reject the null hypothesis when it is false is called a Type I error.

13. The probability of making a Type II error is traditionally called alpha and is represented by α.

14. The probability that an investigator will correctly reject the null hypothesis when it is false is called the power of the statistical test.

15. The value of the alpha level has no effect on the power of a statistical test.

16. The more conservative the alpha level, the less powerful the statistical test will be, everything else being equal.

17. The larger the sample size, the more powerful the statistical test will be, everything else being equal.

18. As a rough guide, investigators generally attempt to achieve statistical power in the range of .05 to .95, depending on the nature of the proposition being investigated.

19. If the null hypothesis is rejected, the results of a statistical test are said to be statistically and practically significant.

20. A statistical analysis that produces a "statistically significant" result may or may not have important practical applications.

21. A directional alternative hypothesis is one that specifies that a population mean is different from a given value and also indicates the direction of that difference.

22. A directional test is one designed to detect differences from a hypothesized population mean score in one direction only.

23. Since they focus on two tails of the distribution, directional tests are often referred to as two-tailed tests.

24. In general, a directional test will always be more powerful than a corresponding nondirectional test if the actual population mean and the hypothesized population mean differ in the specified direction.

25. The t distribution is similar to the normal distribution in that it is bell shaped, unimodal, and symmetrical.

26. As with the normal distribution, there is a single t distribution for every sample size.

27. When $N > 40$, and the raw scores in the population are normally distributed, then the normal and t distributions are very different from one another.

28. A t value is analogous to a z score except that it represents the number of estimated standard errors a sample mean is from μ.

29. The one-sample t test is based on the assumption that the scores on the variable are normally distributed in the population.

30. Statisticians have discovered that the one-sample t test is not robust to violations of the normality assumption under any conditions.

31. The robustness of a test is influenced by several factors, including sample size, the degree of violation, and the form of violation.

32. The confidence interval most commonly used by researchers in the behavioral sciences is the 100% confidence interval.

33. Confidence intervals are calculated around the observed sample mean, $\overline{X}$, rather than μ.

Answers to True-False Items

1. T	11. F	21. T	31. T
2. F	12. F	22. T	32. F
3. T	13. F	23. F	33. T
4. T	14. T	24. T	
5. F	15. F	25. T	
6. T	16. T	26. F	
7. T	17. T	27. F	
8. F	18. F	28. T	
9. T	19. F	29. T	
10. F	20. T	30. F	

Practice Questions: Short Answer

1. Briefly describe the steps involved in the hypothesis testing process.

2. Under what conditions is it appropriate to use the one sample z test when testing hypotheses about the value of a population mean?

3. What is the null hypothesis?

4. Distinguish between a Type I error and a Type II error.

5. What is the power of a statistical test? What factors influence the power of a statistical test?

6. Distinguish between directional and nondirectional tests.

7. Which is more powerful--a directional or a nondirectional test?

8. Describe the similarities and differences between the t distribution and the normal distribution.

9. What are the assumptions of the one-sample t test?

10. What does it mean to say that a test is robust to violations of a distributional assumption?

Answers to Short Answer Questions

1. In the process of hypothesis testing, the investigator begins by stating a proposal, or hypothesis, that is assumed to be true. Based on this assumption, an expected result is specified. The data are collected and the observed result is compared with the expected result. If the observed result is so discrepant from the expected result that the difference cannot be attributed to chance, then the original hypothesis is rejected. Otherwise, it is not rejected.

2. When the variable under study is quantitative in nature and measured on a level that at least approximates interval characteristics, hypotheses about the value of a population mean can be tested using the one sample z test if the population variance is known.

3. The null hypothesis is the hypothesis of "no difference." More technically, it is the hypothesis that we assume to be true for purposes of conducting a statistical test.

4. Once an investigator has drawn a conclusion with respect to the null hypothesis, that conclusion can be either correct or in error. Two types of error are possible. A Type I error involves rejection of the null hypothesis when it is true. A Type II error involves failure to reject the null hypothesis when it is false.

5. $1-\beta$ defines the probability that an investigator will correctly reject the null hypothesis when it is false, and this probability is called the power of the statistical test. The power of the statistical test is influenced by three factors. The alpha level directly affects the power of the statistical test, with more conservative alpha levels yielding less powerful tests, everything else being equal. The power of the statistical test is also influenced by the sample size: The larger the sample size, the more powerful the statistical test will be, everything else being equal. Finally, power is influenced by how discrepant the population mean is from the value of the population mean specified in the null hypothesis. The larger the discrepancy, the more powerful the statistical test will be, everything else being equal.

6. A directional test is one designed to detect differences from a hypothesized population mean score (for example, μ = 100) in one direction only. In contrast, a nondirectional test is one that is designed to detect differences either above or below the hypothesized population mean.

7. In general, a directional test will be more powerful than a corresponding nondirectional test if the actual population mean and the hypothesized population mean differ in the specified direction. However, if the actual population mean differs from the hypothesized population mean in the opposite direction from that stated in the alternative hypothesis, a nondirectional test will be more powerful than its directional counterpart.

8. The t distribution is similar to the normal distribution in that it is bell shaped, unimodal, and symmetrical. In addition, as is the case with a distribution of z scores, the mean of the t distribution is zero. As with the normal distribution, there is not a single t distribution, but rather an entire family of t distributions. However, unlike the normal distribution, the shape of the t distribution is influenced by its degrees of freedom. When the degrees of freedom are relatively large, then the t distribution is very similar to the normal distribution.

9. The one-sample t test is based on the following assumptions: (1) The sample is independently and randomly selected from the population of interest. In most applications, independence is achieved by ensuring that the scores on the variable are provided by different individuals; (2) the scores on the variable are normally distributed in the population. This is known as the normality assumption.

10. When we say that a test is robust to violations of a distributional assumption, we mean that the frequency of Type I and Type II errors, and, thus, the accuracy of our conclusions, are relatively unaffected when the assumption is violated as compared to when the assumption is met.

Answers to Selected Exercises from Textbook

Exercise 26: Exercise 26 tells us to 95% and 99% confidence intervals for the problem in Exercise 20. Exercise 20 provides the sample size N = 30, the sample mean, $\overline{X}$ = 121.00, and the standard deviation estimate, ŝ = 10.77. Our first step is to write out the confidence interval formula to see what we need to compute. Since the standard deviation estimate is given, we use Equation 8.4.

$$CI = \overline{X} - (t)(ŝ_{\overline{X}}) \text{ to } \overline{X} + (t)(ŝ_{\overline{X}})$$

We need to compute the estimated standard error of the mean ($\hat{s}_{\bar{X}}$) and the Appendix D value for t. Since we have been given the standard deviation estimate, we use Equation 7.5 for the estimated standard error of the mean.

$$\hat{s}_{\bar{X}} = \frac{\hat{s}}{\sqrt{N}}$$

$$= \frac{10.77}{\sqrt{30}} = 1.966$$

The test is nondirectional (two-tailed), alpha is .05, and the degrees of freedom are N - 1 = 29. The t values from Appendix D are ±2.045. Now we substitute these numbers into the confidence interval formula:

$$CI = 121.00 - (2.045)(1.966) \text{ to } 121.00 + (2.045)(1.966)$$

$$= 116.98 \text{ to } 125.02.$$

The 95% confidence interval is 116.98 to 125.02.

Exercise 26 also tells us to compute the 99% confidence interval. We can use the computations for the 95% confidence interval, but determine the appropriate value for t. We again use Equation 8.4.

$$CI = \bar{X} - (t)(\hat{s}_{\bar{X}}) \text{ to } \bar{X} + (t)(\hat{s}_{\bar{X}})$$

The test is nondirectional (two-tailed), alpha is .01, and the degrees of freedom are N - 1 = 29. The t values from Appendix D are ±2.756. Now we substitute these numbers into the confidence interval formula.

$$CI = 121.00 - (2.756)(1.966) \text{ to } 121.00 + (2.756)(1.966)$$

$$= 115.58 \text{ to } 126.42.$$

The 99% confidence interval is 115.58 to 126.42.

Exercise 42:   In this exercise, we are told that the fertility rate for zero population growth is 2.11; therefore, if Catholics in the United States are reproducing at a rate that is not different from 2.11, then they are reproducing at zero population growth. Our null hypothesis is that the population mean is 2.11, H_0: $\mu = 2.11$. The exercise tells us to use a nondirectional test, so the alternative hypothesis is that the population mean is not 2.11, H_1: $\mu \neq 2.11$. The traditional alpha level of .05 will be used.

For this problem, we use a one-sample t test. This is because the population standard deviation (σ) is not known. This makes it impossible to compute the standard error of the mean, and therefore we cannot do a one sample z test. Instead, we use the standard deviation estimate to find the estimated standard error of the mean and do a one sample t test. Our first step is to write the formula for the one-sample t test to see what we need to compute (Equation 8.2).

$$t = \frac{\overline{X} - \mu}{\hat{s}_{\overline{X}}}$$

To compute t, we need the estimated standard error of the mean (Equation 7.5).

$$\hat{s}_{\overline{X}} = \frac{\hat{s}}{\sqrt{N}}$$

To compute the estimated standard error of the mean, we need the standard deviation estimate (Equation 7.2), $\hat{s}$. To compute the standard deviation estimate, we need the variance estimate (Equation 7.1).

$$\hat{s}^2 = \frac{SS}{N-1}$$

To compute the variance estimate, we need the sum of squares (Equation 3.6).

$$SS = \Sigma X^2 - \frac{(\Sigma X)^2}{N}$$

Thus, we begin by computing N, ΣX, and ΣX^2, which, for the data in Exercise 42 are N = 25; $\Sigma X = 74$; and $\Sigma X^2 = (4^2 + 5^2 + 2^2 + ... + 2^2 + 2^2) = 298$. Now we substitute these numbers into the appropriate formulas.

$$SS = 298 - \frac{(74)^2}{25} = 298 - 219.04 = 78.96$$

$$\hat{s}^2 = \frac{78.96}{25-1} = 3.29$$

$$\hat{s} = \sqrt{3.29} = 1.814$$

$$\hat{s}_{\bar{X}} = \frac{1.814}{\sqrt{25}} = .363$$

$$t = \frac{2.96 - 2.11}{.363} = 2.34$$

For a nondirectional test, with an alpha level of .05 and degrees of freedom of N - 1 = 24, the critical values of t from Appendix D are ±2.064. Since our observed t of 2.34 exceeds 2.064, we reject the null hypothesis and conclude that Catholics in the United States are, on the average, reproducing at a rate that is different from a zero population growth rate. In fact, since Catholics in our sample have an average of 2.96 children, and zero population growth rate is 2.11, we can state that Catholics in the United States, on the average, are reproducing at a rate that is higher than the rate for zero population growth.

If we write these results using the principles developed in the Method of Presentation section, we obtain (Note: The standard deviation was computed using Equation 7.2 for a standard deviation estimate in Chapter 7):

Results

The mean number of children in the sample (M = 2.96, SD = 1.81) was compared against a hypothesized fertility rate of 2.11 using a one-sample t test, using an alpha level of .05. The difference was statistically significant (t(24) = 2.34, p < .05), indicating that Catholics in the United States are reproducing at a rate above zero population growth rate.

Chapter 9: Research Design and Statistical Preliminaries for Analyzing Bivariate Relationships

Study Objectives

This chapter discusses principles of research design. After reading the material, you should be able to describe observational and experimental research strategies. You should be able to define confounding and disturbance variables and describe methods for controlling them, including matching, holding a variable constant, and random assignment. You should know the difference between a between-subject design and a within-subject design and the relative advantages and disadvantages of each.

You should be able to distinguish parametric and nonparametric statistics and to explain the concept of robustness. You should be able to characterize the assumptions of normality and homogeneity of variance.

Study Tips

Most students readily understand the concept of a confounding variable as discussed in this chapter, but have a more difficult time with the concept of a disturbance variable. You should make an extra effort to follow the logic of disturbance variables. Disturbance variables create variability in the dependent variable and, as we know from Chapter 7, the more variability there is in a variable, the more sampling error there is when estimating parameters with respect to that variable, everything else being equal. By creating more "random noise" in the system, disturbance variables lower statistical power and make it harder to detect effects that are operating.

Glossary Of Important Terms

Study the terms listed below. Make sure you understand each so that you could explain them to someone else who does not know them.

Bivariate Relationship: The relationship between two variables.

Experimental Strategy: When a set of procedures or manipulations is performed to create different values of the independent variable for the research participants.

Observational Strategy: Does not involve the process of actively creating values on an independent variable, but rather involves measuring differences in values that naturally exist in the research participants.

Non-Experimental Strategy: The same as an observational strategy.

Control Group: A group of participants in an experiment that are not exposed to any experimental treatment.

Random Assignment: Assigning people to groups using random procedures so that a person is equally likely to be assigned to any of the experimental or control conditions.

Confounding Variable: A variable that is related to the independent variable (the presumed cause) and that affects the dependent variable (the presumed effect), rendering a causal inference between the independent variable and the dependent variable ambiguous.

Disturbance Variable: A variable that is unrelated to the independent variable (and hence, not confounded with it), but that affects the dependent variable.

Levels: The levels of an independent variable are the different values that the independent variable can have. For example, for gender, there are two levels or values, male and female.

Holding a Variable Constant: A strategy used to control for the effects of a disturbance variable or a confounding variable. It involves selecting research participants in such a way that they all have the same value on the variable being held constant (e.g., for gender, only including females in the study).

Matching: A strategy that is used to control for the adverse effects of confounding variables and disturbance variables. In this approach, an individual in one group is "matched" with and individual in each of the other groups such that all of these individuals have the same value on the confounding/disturbance variable.

Between-Subjects Designs: Research designs involving between-subjects independent variables. A between-subjects independent variable is one where a research participant who has a given value of the independent variable cannot also have a different value on the independent variable. Stated more informally, different individuals are used for each value or level of the independent variable. For example, if the independent variable is gender, then the individuals who have the value "female" are not the same individuals who have the value "male."

Independent Groups Designs: Another word for between-subjects designs.

Within-Subjects Designs: Research designs involving within-subjects independent variables. A within-subjects independent variable is one where the same individual is used at each level or value of the independent variable. For example, if a study is examining the effects of alcohol on learning in a recognition task, the independent variable might be the presence or absence of alcohol. An individual would first perform the recognition task-learning test while not under the influence of alcohol and then again, while under the influence of alcohol.

Matched-Subjects Designs: A strategy where individuals at different levels of the independent variable are matched and then the data are treated as if it is from a within-subjects design.

Correlated Groups Designs: Another word for within-subjects designs.

Repeated Measures Designs: Another word for within-subjects designs.

Carry-Over Effects: An effect that can occur in within-subjects designs. It occurs when the research participant's experience in the first condition s/he affects performance in the second condition.

Parametric Statistics: Techniques that involve the analysis of means, variances, and standard deviations and require quantitative dependent variables. The methods are usually applied when these variables are measured at a level that at least approximates interval characteristics. They make assumptions about the distributions of scores in a population.

Nonparametric Statistics: Class of statistics that tend to focus on medians, interquartile ranges, and ordinal level measures and make no assumptions about the distribution of scores in a population.

Monte Carlo Study: A computer simulation study in which the statistician uses the computer to generate scores for hypothetical populations that he or she knows violates a distributional assumption in some way. Then the test statistic in question is applied many times to random samples from the population to determine the effects of assumption violations on the frequency of Type I and Type II errors.

Cross Sectional Design: A study where age is a between-subjects independent variable.

Longitudinal Design: A study where age is a within-subjects independent variable.

Cohort Effects: Refers to historical variables that are confounded with age and therefore serve as alternative explanations to developmental processes.

Practice Questions: True-False Format

1. When studying the relationship between two variables, the investigator is essentially interested in determining how the different values of one variable are associated with the values of another variable.

2. In an experimental strategy, a set of procedures or manipulations is performed in order to create different values of the dependent variable for the research participants.

3. The advantage of including a control group when utilizing an experimental strategy is that it provides a baseline for evaluating the effects of the experimental manipulation.

4. In contrast to an experimental strategy, an observational strategy or a non-experimental strategy involves the process of actively creating values on an independent variable for the research participants.

5. Experimental strategies automatically control for confounding variables.

6. A major goal of research design is to create alternative explanations.

7. Random assignment is feasible only when an investigator is using a manipulative experimental strategy to "create" values on an independent variable.

8. Random assignment guarantees that the research groups will not differ beforehand on the dependent variable.

9. One way an investigator can reduce sampling error is to reduce the sample sizes for the various groups.

10. A procedure for reducing sampling error is to define the groups such that the variances of scores in the populations (σ^2) will be smaller.

11. A confounding variable is one that is related to the independent variable (the presumed cause) and that affects the dependent variable (the presumed effect), rendering a relational inference between the independent variable and the dependent variable ambiguous.

12. Random assignment is an effective method for controlling confounding variables defined by individual differences.

13. A disturbance variable is one that is related to the independent variable (and hence, confounded with it) but that also affects the dependent variable.

14. A disturbance variable increases sampling error by increasing variability within groups.

15. The major disadvantage of holding a variable constant is that it may restrict the generalizability of the results.

16. With the technique of matching, an individual in one group is "matched" with an individual in each of the other groups such that all of these individuals have the same value on the independent variable.

17. In practice, it is often difficult to identify appropriate variables to serve as a basis for matching, and, once identified, to readily complete the matching process.

18. Since random assignment cannot be applied to experimental groups, experimental variables will always be confounded with all other variables that are naturally related to them.

19. It is entirely possible for two variables to be related to one another, but for no causal relationship to exist between them.

20. The ability to make a causal inference between two variables is a function of the statistical techniques one uses, not the research design used to generate the data that are analyzed by those statistical techniques.

21. Since experimental independent variables will always be confounded with all other variables that are naturally related to them, causal inferences are typically not possible when an experimental research strategy is used.

22. When an experimental research strategy is used, inferences of causation can be made only when confounded variables are controlled.

23. A between-subjects variable is one whose values are "split up" between subjects instead of occurring completely within the same individuals.

24. Between-subjects designs and within-subjects designs are both viable strategies only when the independent variable is experimental in nature.

25. For many observational independent variables, only between-subjects designs are applicable.

26. One advantage of the between-subjects approach as compared with the within-subjects approach is that it is more economical in terms of the number of research participants.

27. The within-subjects design can offer considerably more experimental control than the between-subjects design.

28. One potential problem with between-subjects designs is the fact that the treatment in the first condition may have carry-over effects that influence performance in the second condition.

29. Nonparametric statistics require quantitative dependent variables and are usually applied when these variables are measured on a level that at least approximates interval characteristics.

30. Robustness refers to the extent to which conclusions drawn on the basis of a statistical test (for example, rejection of the null hypothesis) are unaffected by violations of the assumptions underlying the test.

31. The normality and homogeneity of variance assumptions relate only to samples, not to the populations from which the samples were drawn.

32. A Monte Carlo study is a computer simulation study in which the statistician uses the computer to generate scores for hypothetical populations that he or she knows violates a distributional assumption in some way.

Answers to True-False Items

1. T	11. T	21. F	31. F
2. F	12. T	22. T	32. T
3. T	13. F	23. T	
4. F	14. T	24. F	
5. F	15. T	25. T	
6. F	16. F	26. F	
7. T	17. T	27. T	
8. F	18. F	28. F	
9. F	19. T	29. F	
10. T	20. F	30. T	

Practice Questions: Short Answer

1. Distinguish between an experimental strategy and an observational strategy.

2. Describe how random assignment helps to control for alternative explanations.

3. What is a confounding variable and how can they be controlled?

4. What is a disturbance variable and how can they be controlled?

5. What is the major disadvantage of holding a variable constant?

6. What is matching?

7. On what basis are we able to make a causal inference about the relationship between variables?

8. Compare and contrast between- versus within-subjects variables and designs.

9. What are the relative strengths and weaknesses of between-subjects versus within-subjects designs?

10. Distinguish between parametric and nonparametric statistics.

Answers to Short Answer Questions

1. In an experimental strategy, a set of procedures or manipulations is performed in order to create different values of the independent variable for the research participants. An observational strategy or a non-experimental strategy does not involve the process of actively creating values on an independent variable, but rather involves measuring differences in values that naturally exist in the research participants.

2. If the condition in which a given person participates is determined on a completely random basis, it increases the probability that the research groups will not differ beforehand on the dependent variable.

3. A confounding variable is one that is related to the independent variable (the presumed cause) and that affects the dependent variable (the presumed effect), rendering a relational inference between the independent variable and the dependent variable ambiguous. Random assignment is one method for controlling confounding variables defined by individual differences. Holding a variable constant and matching are procedures also used by behavioral scientists to control for confounding variables.

4. A disturbance variable is one that is unrelated to the independent variable (and, hence, not confounded with it) but that affects the dependent variable. As a result, a disturbance variable increases sampling error by increasing variability within groups. Disturbance variables serve to obscure or mask a relationship that exists between the independent and dependent variables. A common strategy used by behavioral scientists to control for disturbance variables is holding a variable constant.

5. The major disadvantage of holding a variable constant is that it may restrict the generalizability of the results. When we hold a variable constant, we have no way of knowing the extent to which the results will generalize across the different levels of the variable that is held constant.

6. In matching, an individual in one group is "matched" with an individual in each of the other groups such that all of these individuals have the same value on the confounding variable.

7. The ability to make a causal inference between two variables is a function of one's research design, not the statistical technique used to analyze the data that are yielded by that research design. Since observational independent variables will always be confounded with all other variables that are naturally related to them, causal inferences are typically not possible when an observational research strategy is used. When an experimental research

strategy is used, inferences of causation can be made only when confounding variables are controlled.

8. A between-subjects variable is one whose values are "split up" between subjects instead of occurring completely within the same individuals. A within-subjects variable is one whose values occur completely within the same individuals. Between-subjects designs and within-subjects designs are both viable strategies when the independent variable is experimental in nature. For many observational independent variables, only between-subjects designs are applicable.

9. One advantage of the within-subjects approach is that it is more economical in terms of research participants. A second advantage of within-subjects designs concerns the control of confounding variables. The within-subjects design can offer considerably more experimental control than the between-subjects design. However, one potential problem with within-subjects designs is the fact that the treatment in the first condition may have carry-over effects that influence performance in the second condition. When carry-over effects are possible, a between-subjects design may be more appropriate. When an investigator is confident that no carry-over effects will occur, a within-subjects design is usually superior to a between-subjects research design.

10. Parametric statistics involve the analysis of means, variances, and standard deviations. Parametric statistics require quantitative dependent variables and are usually applied when these variables are measured on a level that at least approximates interval characteristics. They also require assumptions about the distribution of scores within the populations of interest. In contrast, nonparametric statistics tend to focus on medians, interquartile ranges, and ordinal level measures. In addition, they do not require assumptions about distributional properties of scores that parametric statistics rely on.

Chapter 10: Independent Groups t Test

Study Objectives

This chapter presents the independent groups t test, a test that examines the relationship between two variables. After reading the material in this chapter, you should be able to specify the conditions when you would use the independent groups t test to analyze a bivariate relationship. You should be able to define and characterize a sampling distribution of the difference between two independent means as well as the standard error of the difference between two independent means. You should be able to apply the independent groups t test to determine if a relationship exists between two variables.

You should be able to calculate and interpret eta squared for the independent groups t test. You should be able to define and calculate a sum of squares total, a treatment effect, a grand mean, a sum of squares error and a nullified score.

You should be able to determine the nature of a relationship based on an independent groups t test.

You should be able to specify the assumptions underlying the independent groups t test and briefly characterize the robustness of the test to assumption violations.

You should be able to determine the sample size necessary to achieve a given level (e.g., 0.80) of statistical power when designing a study that will use an independent groups t test. You should be able to explain how alpha, sample size, and the value of eta squared in the population influences statistical power.

You should be able to write up the results of an independent groups t test using APA format as discussed in the Method of Presentation section. You should be able to interpret the results of an independent groups t test from examples reported in the literature.

Study Tips

This is a long chapter because we introduce many concepts that are used in future chapters. The chapter is organized around three major questions to help you keep the "big picture" in mind when analyzing a relationship between an independent variable and a dependent variable. These are (1) is there a relationship between the variables, (2) what is the strength of the relationship, and (3) what is the nature of the relationship? Most of the statistical tests that we consider throughout this book are designed to provide perspectives on these three questions. You should always be relating the material you learn to these questions. Again, try not to get caught up in the computational details too much to the exclusion of a broader conceptual focus. Keep in mind the three questions that you are trying to answer.

A common mistake that students make is to mix up a standard error of the difference between two independent means and the *estimated* standard error of the difference between two independent means. The former is a population parameter whereas the latter is an estimate of the population parameter based on sample data. Be careful to keep this distinction in mind.

Another common error is to misspecify the null and alternative hypotheses as

$$H_0: \overline{X}_1 - \overline{X}_2 = 0$$

$$H_1: \overline{X}_1 - \overline{X}_2 \neq 0$$

This characterization is not correct because it uses sample notation. The null and alternative hypotheses are always stated with respect to population parameters:

$$H_0: \mu_1 - \mu_2 = 0$$

$$H_1: \mu_1 - \mu_2 \neq 0$$

Glossary Of Important Terms

Study the terms listed below. Make sure you understand each so that you could explain them to someone else who does not know them.

Independent Groups t Test: A statistical procedure used to test a null hypothesis of equal population means when the independent variable is between-subjects in nature and has two values and the dependent variable is quantitative.

Sampling Distribution of the Difference Between Two Independent Means: A set of scores that are mean differences between two independent samples for all possible pairs of samples of given sizes.

Standard Error of the Difference: The standard deviation of a sampling distribution of the difference between two independent means. It reflects the average amount of sampling error in a mean difference.

Homogeneity of Variance: The assumption that the population variances are equal.

Pooled Variance Estimate: Combining two sample estimators of the population variance so as to increase the degrees of freedom on which the estimate is based and thereby obtain a better estimate.

Sum of Squares Total: A sum of squares calculated on the dependent variable across all individuals in the experiment.

Grand Mean: The mean value on the dependent variable for all individuals in the study.

Treatment Effect: The mean for a group minus the grand mean.

Nullified Score: A score on the dependent variable in which the effect of the independent variable has been removed.

Unexplained Variance: Variability in the dependent variable that is due to factors other than the independent variable.

Error Variance: Same as unexplained variance.

Sum of Squares Error: A sum of squares calculated on scores on the dependent variable after the effects of the independent variable have been statistically removed (i.e., on nullified scores).

Explained Variance: Variability in the dependent variable that is associated with (explained by) the independent variable.

Sum of Squares Explained: A sum of squares calculated across individuals where an individual's score is the mean score of the condition in which the individual participates.

Eta-Squared: A statistical index that reflects the proportion of variability in the dependent variable that is associated with the independent variable for the sample data.

Levene Test: A statistical procedure for testing a null hypothesis that two population variances are equal.

Practice Questions: True-False Format

1. The independent groups t test is typically used to analyze the relationship between two variables when the independent variable has three or more levels.

2. The independent groups t test is typically used to analyze the relationship between two variables when the dependent variable is quantitative in nature and is measured on a level that at least approximates interval characteristics, and when the independent variable is between-subjects and has two levels.

3. The null hypothesis in the independent groups t test is that a population mean has a specific value, specified by the investigator.

4. In an independent groups t test, the null hypothesis is $H_0: \mu_1 \neq \mu_2$.

5. The alternative hypothesis states that there is a relationship between the independent variable and the dependent variable.

6. The distribution of mean differences across all possible pairs of random samples of a given size represents a sampling distribution of the difference between two independent means.

7. Many properties of a sampling distribution of the mean do not hold for a sampling distribution of the difference between two independent means.

8. The mean of a sampling distribution of the difference between two independent means is always equal to the difference between the population means.

9. The standard deviation of the sampling distribution of the difference between two independent means is called the standard error of the difference between two independent means or, more simply, the standard error of the difference.

10. Like the standard error of the mean, the standard error of the difference is positively skewed.

11. The independent groups t test requires that n_1 equal n_2.

12. Analogous to the standard error of the mean, the size of the standard error of the difference is influenced by two factors: (1) the sample sizes (n_1 and n_2) and (2) the variability of scores in the populations (σ_1^2 and σ_2^2).

13. The standard error of the difference becomes smaller as the difference between the sample means decreases and the variability of scores in the populations increases.

14. The independent groups t test assumes that the two population variances are unequal, or heterogeneous.

15. The quantity σ^2 can best be estimated by combining, or pooling, the variance estimates from the two samples to obtain a pooled variance estimate.

16. The set of all t values more extreme than the critical values constitute an unexpected result or, more formally, the rejection region.

17. The degrees of freedom for the independent groups t test are equal to $n_1 - n_2 + 2$.

18. If we assume the null hypothesis is true in a nondirectional test, then μ_1 and μ_2 will be equal, so $\mu_1 - \mu_2$ will equal 0.

19. When the observed t value is larger than the positive critical t value or smaller than the negative critical t value, this suggests that the observed mean difference can be attributed to sampling error.

20. The t statistic approximately follows a t distribution when (1) the samples are independently and randomly selected from their respective populations, (2) the scores in each population are normally distributed, and (3) the scores in the two populations have equal variances.

21. The independent groups t test is not robust to any violation of the normality and homogeneity of variance assumptions.

22. If we reject the null hypothesis and conclude that a relationship exists between the independent and dependent variables, it is meaningless to ask how strong the relationship is.

23. The sum of squares total represents the total amount of variability that exists in the data.

24. Treatment effects are defined as the difference between a given group mean and the grand mean.

25. Whereas the sum of squares total reflects the total amount of variability in the dependent variable, the sum of squares error reflects the amount of variability that remains before the effects of the dependent variable have been removed.

independent

99

26. Explained variance is variability associated with (explained by) disturbance variables.

27. The total variability in the dependent variable, as represented by SS_{TOTAL}, can be split up, or partitioned, into two components, one ($SS_{EXPLAINED}$) reflecting the influence of the independent variable and one (SS_{ERROR}) reflecting the influence of disturbance variables.

28. The stronger the influence of the independent variable on the dependent variable, the larger the sum of squares error should be relative to the sum of squares total, everything else being equal.

29. The proportion of sample variability in the dependent variable that is explained by the independent variable is called eta-squared.

30. Eta-squared can range from -1.0 to +1.0.

31. 1.00 minus eta-squared represents the proportion of variability in the dependent variable that is not associated with the independent variable--that is, the proportion of sample variability that is due to disturbance variables.

32. Eta-squared is a biased estimator in that it tends to slightly underestimate the strength of the relationship in the population across random samples.

Answers to True-False Items

1. F	11. F	21. F	31. T
2. T	12. T	22. F	32. F
3. F	13. F	23. T	
4. T	14. F	24. T	
5. T	15. T	25. F	
6. T	16. T	26. F	
7. F	17. F	27. T	
8. T	18. T	28. F	
9. T	19. F	29. T	
10. F	20. T	30. F	

Practice Questions: Short Answer

1. When is the independent groups t test used to analyze the relationship between two variables?

2. What is a sampling distribution of the difference between two independent means?

3. What is the standard error of the difference?

4. What factors influence the size of the standard error of the difference?

5. What does a t value represent?

6. What are the assumptions of the independent groups t test?

7. What factors influence the robustness of the independent groups t test?

8. What is eta-squared?

9. What is the rationale for calculating eta-squared following a nonsignificant statistical test?

10. How do we determine what the nature of the relationship is between two variables?

Answers to Short Answer Questions

1. The independent groups t test is typically used to analyze the relationship between two variables when (1) the dependent variable is quantitative in nature and is measured on a level that at least approximates interval characteristics; (2) the independent variable is between-subjects in nature (it can be either qualitative or quantitative); and (3) the independent variable has two and only two levels.

2. A sampling distribution of the difference between two independent means is a theoretical distribution of the differences between two means for all possible random samples of a given size. It is conceptually similar to a sampling distribution of the mean.

3. The standard error of the difference is the standard deviation of the sampling distribution of the difference between two independent means. Like the standard error of the mean, the

standard error of the difference indicates how much sampling error will occur, on the average.

4. Analogous to the standard error of the mean, the size of the standard error of the difference is influenced by two factors: (1) the sample sizes (n_1 and n_2) and (2) the variability of scores in the populations (σ_1^2 and σ_2^2). The standard error of the difference becomes smaller as the sample sizes increase and the variability of scores in the populations decreases.

5. A t value represents how many estimated standard errors the observed difference between sample means is away from the hypothesized difference between the population means as stated in the null hypothesis.

6. The t statistic approximately follows a t distribution when the following assumptions are met: (1) the samples are independently and randomly selected from their respective populations, (2) the scores in each population are normally distributed, and (3) the scores in the two populations have equal variances; that is, $\sigma_1^2 = \sigma_2^2$.

7. For the t test to be valid, it is important that the assumption of independent and random selection be met. However, under certain conditions, the independent groups t test is robust to violations of the normality and homogeneity of variance assumptions. If the sample sizes in the two groups are each greater than 40 and roughly comparable, then the test is robust to rather severe departures of the normality assumption. When the sample sizes are equal, the t test is quite robust to violations of the assumption of homogeneity of variance. This also holds for unequal n in the two groups, as long as the sample sizes are comparable. With fairly discrepant sample sizes, there is a tendency for the test to be conservative when the group with the larger sample size has the larger variance. By the same token, the t test is liberal when the larger sample size is paired with the smaller variances.

8. Eta-squared indexes the strength of the relationship between the independent and dependent variables. It represents the proportion of variability in the dependent variable that is associated with the independent variable.

9. A large eta squared coupled with a statistically non-significant t value is often suggestive of low statistical power.

10. The nature of the relationship between two variables is determined by examining the sample means. The conclusion is intended to apply to the population means.

Answers to Selected Exercises from Textbook

Exercise 4: In this exercise, we are asked to compute the pooled variance estimate and estimated standard error of the difference. The exercise provides the sample size for each group and the variance estimate for each group.

$$n_1 = 10 \qquad n_2 = 13$$

$$\hat{s}_1^2 = 6.48 \qquad \hat{s}_2^2 = 4.73$$

We use Equation 10.3 for the pooled variance estimate. First, write out the formula so we can see what we need to compute. The numbers we need have been provided, so we simply substitute them into the formula.

$$\hat{s}^2_{pooled} = \frac{(df_1)(\hat{s}_1^2) + (df_2)(\hat{s}_2^2)}{df_{TOTAL}}$$

$$\hat{s}^2_{pooled} = \frac{(10-1)(6.48) + (13-1)(4.73)}{(10 + 13 - 2)} = \frac{58.32 + 56.76}{21} = 5.48$$

Since we now have the pooled variance estimate, we can use Equation 10.4 to compute the estimated standard error of the difference. First, write the formula, and then substitute the appropriate numbers into the formula.

$$\hat{s}_{\overline{X}1-\overline{X}2} = \sqrt{\frac{\hat{s}^2_{pooled}}{n_1} + \frac{\hat{s}^2_{pooled}}{n_2}}$$

$$= \sqrt{\frac{5.48}{10} + \frac{5.48}{13}} = .98$$

Exercise 11: This exercise asks for the sum of squares total. We use the standard formula for the sum of squares, and compute the sum of squares for the dependent variable across all individuals in the study. First, write the formula so we know what we need to compute (Equation 3.6).

$$SS_{TOTAL} = \Sigma X^2 - \frac{(\Sigma X)^2}{N}$$

We need to calculate N, ΣX^2 and ΣX:

Individual	Gender	Attitude Score (X)	X^2
1	Man	7	49
2	Man	7	49
3	Man	8	64
4	Man	7	49
5	Man	6	36
6	Woman	4	16
7	Woman	3	9
8	Woman	4	16
9	Woman	5	25
10	Woman	4	16
		$\Sigma X = 55$	$\Sigma X^2 = 329$

Next, substitute in the values:

$$SS_{TOTAL} = 329 - \frac{(55)^2}{10} = 26.50$$

The total amount of variability that exists in the data is represented by the sum of squares total, which is 26.50.

Exercise 12: This exercise requires that we compute the treatment effect for men and the treatment effect for women for the same data. Our first step is to write the formulas so we know what we need to compute. A treatment effect is the difference between a given group mean and the grand mean.

$$T_M = \overline{X}_M - G$$

$$T_F = \overline{X}_F - G$$

We need the grand mean and the mean for men and the mean for women.

$$\text{grand mean:} \quad G = \frac{\Sigma X}{N}$$

$$= \frac{55}{10} = 5.50$$

$$\overline{X}_M = \frac{\Sigma X_M}{n_M}$$

$$= \frac{35}{5} = 7.00$$

$$\overline{X}_F = \frac{\Sigma X_F}{n_F}$$

$$= \frac{20}{5} = 4.00$$

Now we substitute these numbers into the formula.

$$T_M = 7.00 - 5.50 = 1.50$$

$$T_F = 4.00 - 5.50 = -1.50$$

The effect of gender was to raise the tendency to hold discriminatory attitudes 1.50 units above the grand mean for men, and to lower the tendency 1.50 units below the grand mean for women.

Exercise 13: This exercise requires us to use the variance extraction procedures to nullify the effect of gender from the dependent variable. We will *subtract* the treatment effect for a given group from each score in the group. The treatment effect is 1.50 for men and -1.50 for women. This yields the following:

Gender	Attitude Scale (X)	Treatment Effect (T)	Nullified Score ($X_n = X - T$)
Man	7	1.50	5.50
Man	7	1.50	5.50
Man	8	1.50	6.50
Man	7	1.50	5.50
Man	6	1.50	4.50
Woman	4	-1.50	5.50
Woman	3	-1.50	4.50
Woman	4	-1.50	5.50
Woman	5	-1.50	6.50
Woman	4	-1.50	5.50

Exercise 14: Exercise 14 asks for the sum of squares error and the sum of squares explained for the data in exercise 13. The sum of squares for the nullified scores will give us the sum of squares error. We simply apply the standard sum of squares formula (Equation 3.6) to the nullified scores. First, write the formula to see what we need to compute.

$$SS_{ERROR} = \Sigma X_n^2 - \frac{(\Sigma X_n)^2}{N}$$

We need ΣX_n^2, ΣX_n, and N:

Score	X_n	X_n^2
1	5.50	30.25
2	5.50	30.25
3	6.50	42.25
4	5.50	30.25
5	4.50	20.25
6	5.50	30.25
7	4.50	20.25
8	5.50	30.25

Score	X_n	X_n^2
9	6.50	42.25
10	5.50	30.25

$$\Sigma X_n = 55.00 \qquad \Sigma X_n^2 = 306.50$$

$$SS_{ERROR} = \Sigma X_n^2 - \frac{(\Sigma X_n)^2}{N}$$

$$= 306.50 - \frac{(55.00)^2}{10}$$

$$= 306.50 - 302.50 = 4.00$$

The index of how much error variance remains after we have removed the influence of the independent variable is the sum of squares error and it is 4.00 for this example.

Exercise 14 also asks for the sum of squares explained for the data in Exercise 13. The sum of squares explained is the variability in the dependent variable associated with the independent variable. We can compute the sum of squares explained by simply subtracting the sum of squares error from the sum of squares total.

$$SS_{EXPLAINED} = SS_{TOTAL} - SS_{ERROR}$$

$$= 26.50 - 4.00 = 22.50$$

The variability associated with the influence of the independent variable on the dependent variable is 22.50.

Exercise 47: In this exercise, the independent variable is the type of environment and it has two levels: enriched or isolated. The dependent variable is the weight of the cortex (the outer layers of the brain) measured in milligrams (mg). Two separate groups of rats were used: one for the enriched environment and one for the isolated environment. The dependent variable is quantitative and measured on approximately an interval level. The independent variable is between-subjects and has two and only two levels. Consequently, we use an independent groups t test.

Exercise 47 asks us to analyze the data and draw a conclusion. We need several pieces of information. After making sure you understand the nature of the study and which statistic to use, begin by stating the hypotheses. (E stands for enriched environment, group 1, and I stands for isolated environment, group 2).

$$H_0: \mu_E = \mu_I$$

$$H_1: \mu_E \neq \mu_I$$

The null hypothesis states that there is no relationship between type of environment and weight of the cortex. The exercise tells us to use a nondirectional test, so the alternative hypothesis states that there is a relationship.

For the t test, first write the formula to see what we need to compute (Equation 10.8).

$$t = \frac{(\overline{X}_1 - \overline{X}_2)}{\hat{s}_{\overline{X}1 - \overline{X}2}}$$

We need to compute the sample means for each group, and the estimated standard error of the difference. Write the formula for the estimated standard error of the difference to see what we need to compute. Since we have raw data, we use Equation 10.6.

$$\hat{s}_{\overline{X}1 - \overline{X}2} = \sqrt{\left(\frac{SS_1 + SS_2}{n_1 + n_2 - 2}\right)\left(\frac{1}{n_1} + \frac{1}{n_2}\right)}$$

109

So, we need the sum of squares for each group. We use the standard formula for the sum of squares; therefore, we need the ΣX, ΣX^2, and n for each group. Let's prepare the data.

Enriched Environment		Isolated Environment	
X_1	X_1^2	X_2	X_2^2
685	469,225	660	435,600
690	476,100	642	412,164
675	455,625	640	409,600
660	435,600	626	391,876
645	416,025	612	374,544
630	396,900	610	372,100
635	403,225	592	350,464

$\Sigma X_1 = 4,620 \qquad \Sigma X_1^2 = 3,052,700$ $\qquad\qquad \Sigma X_2 = 4,382 \quad \Sigma X_2^2 = 2,746,348$

Now substitute the appropriate values into the standard sum of squares formula (Equation 3.6).

$$SS_1 = \Sigma X_1^2 - \frac{(\Sigma X_1)^2}{n_1} \qquad\qquad SS_2 = \Sigma X_2^2 - \frac{(\Sigma X_2)^2}{n_2}$$

$$= 3,052,700 - \frac{(4,620)^2}{7} \qquad\qquad = 2,746,348 - \frac{(4,382)^2}{7}$$

$$= 3,500 \qquad\qquad\qquad\qquad = 3,216$$

110

We are ready to substitute the sum of squares and n into the estimated standard error of the difference formula (Equation 10.6).

$$\hat{s}_{\bar{X}_1 - \bar{X}_2} = \sqrt{\left(\frac{SS_1 + SS_2}{n_1 + n_2 - 2} \right) \left(\frac{1}{n_1} + \frac{1}{n_2} \right)}$$

$$\hat{s}_{\bar{X}_1 - \bar{X}_2} = \sqrt{\left(\frac{3,500 + 3,216}{7 + 7 - 2} \right) \left(\frac{1}{7} + \frac{1}{7} \right)}$$

$$= \sqrt{(559.667)(.286)} = 12.652$$

Now we can compute the t test (Equation 10.8).

$$t = \frac{(\bar{X}_1 - \bar{X}_2)}{\hat{s}_{\bar{X}_1 - \bar{X}_2}}$$

$$= \frac{660 - 626}{12.652} = 2.69$$

The observed t is 2.69. To draw a conclusion, we must test the t for statistical significance. We need the critical values from Appendix D. To use Appendix D, we need directionality, the alpha level, and the degrees of freedom. The test is nondirectional. We adopt an alpha level of .05, since there is no reason to adopt a different level. the degrees of freedom are $n_1 + n_2 - 2 = 12$. Therefore, the critical values from Appendix D are ±2.179. An unexpected result is defined as all t scores less than -2.179 or greater than +2.179. The observed t of 2.69

is greater than the positive critical value of 2.179. We reject the null hypothesis and conclude that there is a relationship between type of environment and the weight of the cortex.

The strength of the relationship is indexed by eta-squared. We need only substitute the appropriate numbers into Equation 10.11.

$$\text{eta}^2 = \frac{t^2}{t^2 + df}$$

$$= \frac{2.69^2}{2.69^2 + 12}$$

$$= \frac{7.236}{19.236} = .38$$

The proportion of variability in weight of the cortex that is associated with the type of environment is .38, which is a strong effect.

The mean of the enriched environment group (660.00) is higher than the mean of the isolated environment group (626.00); therefore, we conclude that an enriched environment produces a heavier cortex than an isolated environment.

Exercise 47 also asks us to write the results using the principles discussed in the Method of Presentation section (Note: All standard deviations were computed using Equation 7.2 for a standard deviation estimate in Chapter 7):

Results

    An independent groups t test compared the mean cortex

weights of rats raised in enriched versus impoverished

environments, using an alpha level of .05. The results

indicated that the cortex weight of rats tended to be greater

when they are raised in an enriched environment ($\underline{M}$ = 660.00 mg,

$\underline{SD}$ = 24.15) than when they are raised in an isolated

environment ($\underline{M}$ = 626.00 mg, $\underline{SD}$ = 23.15), $\underline{t}$(12) = 2.69, $\underline{p}$ < .02.

As indicated by eta-squared, the strength of the relationship

between the type of environment and weight of cortex was .38.

Chapter 11: Correlated Groups t Test

Study Objectives

This chapter presents the correlated groups t test, a test that examines the relationship between two variables. After reading the material in this chapter, you should be able to specify the conditions when you would use the correlated groups t test to analyze a bivariate relationship. You should be able to define and characterize a sampling distribution of the mean of difference scores as well as the standard error of the mean of difference scores. You should be able to apply the correlated groups t test to determine if a relationship exists between two variables.

You should be able to calculate and interpret eta squared for the correlated groups t test and you should be able to determine the nature of a relationship.

You should be able to specify the assumptions underlying the correlated groups t test and briefly characterize the robustness of the test to assumption violations.

You should be able to determine the sample size necessary to achieve a given level (e.g., 0.80) of statistical power when designing a study that will use a correlated groups t test. You should be able to explain how alpha, sample size, and the value of eta squared in the population influences statistical power.

You should understand how the effects of individual differences are "removed" from the dependent variable using nullified scores.

You should be able to discuss the relative advantages and disadvantages of a correlated groups t test as compared with an independent groups t test.

You should be able to write up the results of a correlated groups t test using APA format as discussed in the Method of Presentation section. You should be able to interpret the results of a correlated groups t test from examples reported in the literature.

Study Tips

At this point, a large number of terms have been introduced with the phrase "standard" in them. These include a standard deviation, standard score, standard normal distribution, standard deviation estimate, standard error of the mean, estimated standard error of the mean, standard error of the difference between two independent means, estimated standard error of the difference between two independent means, standard error of the mean of difference scores, and the estimated standard error of the mean of difference scores. Students often confuse these terms. Each refers to a distinct concept. Be sure you can articulate the differences.

It is very important that you understand the conditions where you would apply the correlated groups t test as opposed to the independent groups t test. Make a special effort to understand the relevant distinctions.

Glossary Of Important Terms

Study the terms listed below. Make sure you understand each so that you could explain them to someone else who does not know them.

Correlated Groups t Test: A statistical procedure used to test a null hypothesis of equal population means when the independent variable is within-subjects in nature and has two values and the dependent variable is quantitative.

Sampling Distribution of the Mean of Difference Scores: A theoretical distribution consisting of mean difference scores across individuals for all possible random samples of a given size that could be selected from a population.

Counterbalancing: Using more than one sequence of conditions or levels of the independent variable so as evaluate the effects of order of condition.

Practice Questions: True-False Format

1. The correlated groups t test is typically used to analyze the relationship between two variables when the independent variable is between-subjects in nature.

2. The major difference between the correlated groups t test and the independent groups t test is that the former is used when the independent variable is manipulated and the latter is used when the independent variable is observed.

3. An important advantage of the correlated groups t test over the independent groups t test relates to the control of confounding variables.

4. One major source of "noise" in behavioral science data is the difference in backgrounds and abilities of the individuals participating in an investigation.

5. The correlated groups t test provides a less sensitive test of the relationship between the independent and dependent variables than does the independent groups t test.

6. A test's sensitivity can be defined as its ability to detect a relationship between variables when a relationship exists in the population.

7. A sampling distribution of the mean of difference scores can be defined as a theoretical distribution consisting of mean difference scores across individuals for all possible random samples of a given size that could be selected from a population.

8. The standard deviation of the sampling distribution of mean difference scores is referred to as the standard error of the difference.

9. The mean of the sampling distribution of mean difference scores is equal to the mean difference score across individuals in the population.

10. The correlated groups t test has $N - 2$ degrees of freedom associated with it.

11. One of the underlying assumptions of the correlated groups t test is that the population of mean scores is normally distributed.

12. The assumption of independent, random sampling is not an important one for the correlated groups t test.

13. For sample sizes of 40 or more, the correlated groups t test is quite robust to violations of the normality assumption, even for distributions exhibiting considerable skewness.

14. If the sample size is less than 15, the correlated groups t test may show inflated Type I errors for data that are markedly skewed.

15. The formula for computing eta-squared from t and its degrees of freedom for the correlated groups t test is the same as that for the independent groups t test.

16. The interpretation of eta-squared for the correlated groups t test is the same as that for the independent groups t test.

17. In the correlated groups t test, eta-squared represents the proportion of variability in the dependent variable that is associated with the independent variable after variability due to individual differences has been included.

18. There is less variability in a set of nullified scores because the effects of individual differences in background have been removed.

19. If we were to compute the average nullified score for each research participant, we would find that every person would have the same average score.

20. A correlated groups t test is analogous to an independent groups t test with the effects of individual differences added to the dependent variable.

21. Because the correlated groups t test involves the analysis of nullified scores, the question of the nature of the relationship is not meaningful.

22. The analysis of the nature of the relationship for the correlated groups t test involves examination of the mean scores for the two conditions.

23. One procedure for evenly distributing carry-over effects across conditions is called counterbalancing.

24. With smaller sample sizes, larger population effect sizes are necessary to obtain adequate statistical power.

25. It is possible to obtain a large value of eta-squared without obtaining a statistically significant t value.

26. Unlike the independent groups t test, larger sample sizes will not increase the power of the correlated groups t test.

27. Since variability due to individual differences is extracted from the dependent variable as part of the correlated groups t test procedure, a correlated groups t test will usually be more powerful than a corresponding independent groups t test.

28. It is typically the case that the estimated standard error for the correlated groups t test is larger than the estimated standard error for the independent groups t test.

Answers to True-False Items

1. F	11. F	21. F
2. F	12. F	22. T
3. F	13. T	23. T
4. T	14. T	24. F
5. F	15. T	25. T
6. T	16. F	26. F
7. T	17. F	27. T
8. F	18. T	28. F
9. T	19. T	
10. F	20. F	

Practice Questions: Short Answer

1. When is the correlated groups t test used to analyze the relationship between two variables?

2. What is meant by the sensitivity of a statistical test?

3. What is a sampling distribution of the mean of difference scores?

4. What are the assumptions of the correlated groups t test?

5. Discuss the robustness of the correlated groups t test.

6. What is the difference between eta-squared for the correlated groups t test and eta-squared for the independent groups t test?

7. In what sense are the correlated groups t test and the independent groups t test equivalent?

8. Explain why it is possible to observe a strong relationship between two variables based on the value of eta-squared without obtaining a statistically significant t value.

9. Why is the correlated groups t test usually more powerful than the independent groups t test?

Answers to Short Answer Questions

1. The correlated groups t test is typically used to analyze the relationship between two variables when (1) the dependent variable is quantitative in nature and is measured on a level that at least approximates interval characteristics; (2) the independent variable is within-subjects in nature (it can be either qualitative or quantitative); and (3) the independent variable has two and only two levels.

2. The sensitivity of a statistical test can be defined as its ability to detect a relationship between variables when a relationship exists in the population.

3. A sampling distribution of the mean of difference scores is a theoretical distribution consisting of mean difference scores across individuals for all possible random samples of a given size that could be selected from a population.

4. The assumptions are (1) the sample is independently and randomly selected from the population of interest; (2) the population of difference scores is normally distributed.

5. For the correlated groups t test, the assumption of independent, random sampling is an important one. By contrast, the test is relatively robust to violations of the normality assumption. If the sample size is less than 15, the test may show inflated Type I errors for data that are markedly skewed. However, for sample sizes of 40 or more, the test is quite robust, even for distributions exhibiting considerable skewness.

6. Whereas eta-squared represents the proportion of variability in the dependent variable that is associated with the independent variable in the independent groups case, in the correlated groups case eta-squared represents the proportion of variability in the dependent variable that is associated with the independent variable after the variability due to individual differences has been removed.

7. A correlated groups t test is analogous to an independent groups t test with the effects of individual differences extracted from the dependent variable.

8. With small sample sizes, sampling error will be large and even strong effects may go undetected as a result of lower statistical power.

9. Since variability due to individual differences is extracted from the dependent variable as part of the correlated groups t test procedure, a correlated groups t test will usually (but not always) be more powerful than a corresponding independent groups t test, due to a smaller standard error.

Answers to Selected Exercises from Textbook

Exercise 15: Exercise 15 tells us to extract the effects of individual differences from a set of scores; that is, we are to compute the nullified scores. Then, we are to compute the means for the original scores for both time 1 and time 2, and the means for the nullified scores for both time 1 and time 2.

First, we need to compute the grand mean (G) by summing all the scores in the study and dividing by the number of scores we summed.

$$\text{grand mean:} \quad G = \frac{\Sigma X}{N}$$

$$= \frac{130}{10} = 13.00$$

The average score across subjects and across both conditions (G) is 13.00. Now, we compute the average score for each subject across both conditions, and enter those numbers in the right hand column below.

Subject	X for Time 1	X for Time 2	$\overline{X}_i$
1	10	12	11.00
2	13	17	15.00
3	12	14	13.00
4	11	13	12.00
5	14	14	14.00

The numbers under the right-hand column represent each subject's average score across the two conditions. Now, we subtract the grand mean from each of these scores in the right most column to obtain an average index of individual differences. This subtraction appears as follows:

Subject	$\overline{X}_i$	$(\overline{X}_i - G)$
1	11.00	11.00 - 13.00 = -2.00
2	15.00	15.00 - 13.00 = 2.00
3	13.00	13.00 - 13.00 = 0.00
4	12.00	12.00 - 13.00 = -1.00
5	14.00	14.00 - 13.00 = 1.00

We now subtract this score for each individual from their time 1 score and from their time 2 score. These are the nullified scores, that is, the scores from which the average effect of individual differences has been removed. The nullified scores are the two right-most columns in the table below:

Subject	X for Time 1	X for Time 2	$\overline{X}_i$	Nullified X for Time 1	Nullified X for Time 2
1	10	12	11.00	12.00	14.00
2	13	17	15.00	11.00	15.00
3	12	14	13.00	12.00	14.00
4	11	13	12.00	12.00	14.00
5	14	14	14.00	13.00	13.00
Mean:	12.00	14.00	13.00	12.00	14.00

As you can see, the mean values in each condition are the same for the original scores and the nullified scores. Why is this the case? With the nullified scores, we have extracted the effect of the individual differences, which take the role of disturbance variables. Since disturbance variables are unrelated to the independent variable, the means will not be affected.

Exercise 16: This exercise directs us to conduct a nondirectional independent groups t test on the nullified scores we computed in Exercise 15. First, write the formula for the independent groups t test to see what we need to compute (Equation 10.8).

$$t = \frac{(\overline{X}_1 - \overline{X}_2)}{\hat{s}_{\overline{X}1-\overline{X}2}}$$

To conduct the t test, we need to calculate the estimated standard error of the mean difference. We can use Equation 10.6, but with a minor adjustment. The formula for the independent groups t test is

$$\hat{s}_{\overline{X}1-\overline{X}2} = \sqrt{\left(\frac{SS_1 + SS_2}{df}\right)\left(\frac{1}{n_1} + \frac{1}{n_2}\right)}$$

where df is the degrees of freedom and equals $n_1 + n_2 - 2$. However, for the correlated groups t test, the degrees of freedom is N - 1 (where $N = n_1 = n_2$). This yields

$$\hat{s}_{\overline{X}1-\overline{X}2} = \sqrt{\left(\frac{SS_1 + SS_2}{N - 1}\right)\left(\frac{1}{n_1} + \frac{1}{n_2}\right)}$$

To compute the estimated standard error of the mean difference, we need the sum of squares for each condition. We use the standard sum of squares formula for each condition (Equation 3.6).

$$SS_1 = \Sigma X_1^2 - \frac{(\Sigma X_1)^2}{N}$$

122

$$SS_2 = \Sigma X_2^2 - \frac{(\Sigma X_2)^2}{N}$$

Therefore, we need to compute ΣX and ΣX^2 and N for each condition.

Individual	Nullified data for Time 1 X_1	X_1^2	Nullified data for Time 2 X_2	X_2^2
1	12.00	144.00	14.00	196.00
2	11.00	121.00	15.00	225.00
3	12.00	144.00	14.00	196.00
4	12.00	144.00	14.00	196.00
5	13.00	169.00	13.00	169.00
	$\Sigma X_1 = 60.00$	$\Sigma X_1^2 = 722.00$	$\Sigma X_2 = 70.00$	$\Sigma X_2^2 = 982.00$

Now we substitute the numbers into the appropriate formulas.

$$SS_1 = \Sigma X_1^2 - \frac{(\Sigma X_1)^2}{N} \qquad\qquad SS_2 = \Sigma X_2^2 - \frac{(\Sigma X_2)^2}{N}$$

$$= 722.00 - \frac{(60.00)^2}{5} = 2 \qquad\qquad = 982.00 - \frac{(70.00)^2}{5} = 2$$

$$\hat{s}_{\overline{X}1 - \overline{X}2} = \sqrt{\left(\frac{2 + 2}{5 - 1} \right)\left(\frac{1}{5} + \frac{1}{5} \right)}$$

$$= \sqrt{(1.000)(.400)} = .632$$

$$t = \frac{(\overline{X}_1 - \overline{X}_2)}{\hat{s}_{\overline{X}_1 - \overline{X}_2}}$$

$$t = \frac{12.00 - 14.00}{.632} = -3.16$$

We are to compare the findings here to those for Exercise 12. We are comparing the t test computed on the nullified scores (Exercise 16) to the t test computed on the original scores (Exercise 12). The observed t using the procedure in Appendix 11.1 is -3.16. This is the same value as was obtained in Exercise 12.

Exercise 38: We are interested in the effects of time of assessment of products on ratings of product durability. The independent variable is time of assessment. It has two levels: ratings before making a choice between the two products, and ratings after making a choice between the two products. The dependent variable is the ratings of product desirability, with 1 representing low desirability and 8 representing high desirability.

Since the same subjects were used for both levels of the independent variable, the design of the study is within-subjects. The independent variable has two and only two levels. The dependent variable is quantitative and is measured on approximately an interval level. Thus, a correlated groups t test is appropriate.

Exercise 38 tells us to analyze the data and draw a conclusion. The null hypothesis states that there is no relationship between the independent and dependent variables. The alternative hypothesis is that there is a relationship between the independent and dependent variables, that the effects of the two conditions are not the same.

$$H_0: \mu_B = \mu_A$$

$$H_1: \mu_B \neq \mu_A$$

To begin the analysis, write the formula for the correlated groups t test so we know what we need to compute. Since we have raw data, we use Equation 11.3.

$$t = \frac{\overline{D}}{\hat{s}_{\overline{D}}}$$

To conduct the correlated groups t test, we need the mean difference and the estimated standard error of the mean difference (Equation 11.1):

$$\hat{s}_{\overline{D}} = \frac{\hat{s}_D}{\sqrt{N}}$$

To compute the mean difference, we need the sum of the difference scores and N. To compute the estimated standard error of the mean difference, we need the standard deviation estimate for the sample difference scores.

$$\hat{s}_D = \sqrt{\hat{s}_{\overline{D}}^{\,2}}$$

For the standard deviation estimate, we need the variance estimate of the sample difference scores.

$$\hat{s}_D^{\,2} = \frac{SS_D}{N-1}$$

For the variance estimate, we need the sum of squares of the sample difference scores.

$$SS_D = \Sigma D^2 - \frac{(\Sigma D)^2}{N}$$

Thus, we need to compute ΣD^2, ΣD, and N.

Individual	X for Before Choice	X for After Choice	Difference (D)	D^2
1	8	4	4	16
2	7	3	4	16
3	7	5	2	4
4	4	6	-2	4
5	8	4	4	16
6	6	4	2	4
7	6	2	4	16
8	5	5	0	0
9	5	3	2	4
10	4	4	0	0

$$\Sigma X_B = 6 \qquad \Sigma X_A = 40 \qquad \Sigma D = 20 \qquad \Sigma D^2 = 80$$
$$\overline{X}_B = 6.00 \qquad \overline{X}_A = 6.00 \qquad \overline{D} = 2.00$$

Now we substitute the numbers into the appropriate formulas.

$$SS_D = 80 - \frac{(20)^2}{10} = 40$$

$$\hat{s}_D{}^2 = \frac{40}{10-1} = 4.444$$

$$\hat{s}_D = \sqrt{4.444} = 2.108$$

$$\hat{s}_{\bar{D}} = \frac{2.108}{\sqrt{10}} = .667$$

$$t = \frac{2.00}{.667} = 3.00$$

For a nondirectional test, an alpha level of .05, and degrees of freedom of N - 1 = 9, the critical values of t from Appendix D are ±2.262. Since our observed t of 3.00 exceeds the positive critical value of 2.262, we reject the null hypothesis and conclude that there is a relationship between the independent variable and the dependent variable; that is, between the time of assessment of the product and the ratings of product desirability.

We compute eta-squared to analyze the strength of the relationship (Equation 10.11).

$$eta^2 = \frac{t^2}{t^2 + df}$$

$$= \frac{3.00^2}{3.00^2 + 9}$$

$$= \frac{9.00}{18.00} = .50$$

The eta-squared of .50 indicates that there is a strong relationship between the independent and dependent variables.

Exercise 38 also asks us to write the results using the principles discussed in the Method of Presentation section (Note: All standard deviations were computed using Equation 7.2 for a standard deviation estimate in Chapter 7):

127

Results

The mean desirability ratings for the unchosen alternative before and after making the choice between the two products were compared using a correlated groups t test, using an alpha level of .05. The t test showed that the alternative in question was rated significantly more desirable on the first occasion ($\underline{M}$ = 6.00, $\underline{SD}$ = 1.49) than on the second occasion ($\underline{M}$ = 4.00, $\underline{SD}$ = 1.15), $\underline{t}$(9) = 3.00, $\underline{p}$ < .02. The strength of the relationship between the time of assessment and product desirability was .50, as indexed by eta-squared.

Chapter 12: One-Way Between-Subjects Analysis of Variance

Study Objectives

This chapter presents one-way between-subjects analysis of variance, a test that examines the relationship between two variables. After reading the material in this chapter, you should be able to specify the conditions when you would use one-way between-subjects analysis of variance to analyze a bivariate relationship. You should be able to define and characterize between group variability and within group variability, and the statistical measures of them: The sum of squares between, the mean square between, the sum of squares within, and the mean square within. You should be able to describe the logic of the F test. You should be able to explain what a sampling distribution of an F ratio is and how it relates to the theoretical F distribution. You should be able to apply one-way between-subjects analysis of variance in its entirety to determine if a relationship exists between two variables. This includes knowing how to compute all of the intermediate statistics and forming a summary table for the calculations. You should be able to specify the assumptions underlying one-way between-subjects analysis of variance and briefly characterize the robustness of the test to assumption violations.

You should be able to calculate and interpret eta squared for one-way between-subjects analysis of variance. You should be able to determine the nature of a relationship, using the Tukey HSD test. You should be able to explain why the Tukey HSD test is used instead of multiple t tests.

You should be able to determine the sample size necessary to achieve a given level (e.g., 0.80) of statistical power when designing a study that will use one-way between-subjects analysis of variance. You should be able to explain how alpha, sample size, and the value of eta squared in the population influences statistical power for this test.

You should be able to write up the results of an one-way between-subjects analysis of variance using APA format as discussed in the Method of Presentation section. You should be able to interpret the results of a one-way between-subjects analysis of variance from examples reported in the literature.

Study Tips

This chapter involves many computational steps and, as before, it is easy to lose the big picture of what the test is trying to accomplish. Keep in mind the three questions that we are trying to answer, (1) is there a relationship between the independent and dependent variables (2) what is the strength of the relationship, and (3) what is the nature of the relationship. The first question is addressed by the F test, the second question by eta squared, and the third question by Tukey's HSD test. All of the computations -- the sum of squares between, the degrees of freedom between, the sum of squares within, the degrees of freedom within, the means square between and the mean square within -- are merely steps along the way to getting the F ratio used for purposes of executing the F test. Students easily become confused about all the different sum of squares and mean squares in this chapter. Make a special effort to keep the distinctions between them straight, as this is very important.

Glossary Of Important Terms

Study the terms listed below. Make sure you understand each so that you could explain them to someone else who does not know them.

One-Way Between-Subjects Analysis of Variance: A statistical technique that is used to analyze the relationship between two variables, under the same circumstances as the independent groups t test except that the independent variable has more than two levels. The procedure is focused on testing the null hypothesis of equivalent population means across the levels of the independent variable.

Between-Group Variability: The extent to which mean scores in the different groups defined by the levels of the independent variable are similar.

Within-Group Variability: The variability of scores within a given group.

Variance Ratio: A ratio of two variance estimates.

Partitioning of Variance: Decomposing the total variability into different components or sources. The sum of squares total can be broken down into two parts, the sum of squares between and the sum of squares within.

Sum of Squares Total: A sum of squares as applied to the dependent variable across all individuals in the study.

Sum of Squares Between: A sum of squares as applied to a set of scores for each individual in the experiment, where a given individual's "score" is the mean value of the group that the individual is a member of. It reflects between-group variability.

Sum of Squares Within: A sum of squares as applied to deviation scores from the group mean across all individuals in the study. It reflects within-group variability.

Mean Square Between: The sum of squares between divided by the degrees of freedom between. It is a variance estimate and reflects between-group variability.

Mean Square Within: The sum of squares within divided by the degrees of freedom within. It is a variance estimate and reflects within group variability.

F Ratio: A variance ratio named for the statistician Sir Ronald Fisher. In the present chapter it is the ratio of the mean square between divided by the mean square within.

Sampling Distribution of the F Ratio: A distribution of F ratios calculated on all possible random samples of a given size based on the levels of the independent variable.

F Distribution: A theoretical distribution that has a known form and for which probability statements can be made about the occurrence of different ranges of scores.

F Test: The evaluation of the F ratio formed by dividing the mean square between by the mean square within to determine if it is larger than the critical value of F.

Summary Table: A table that summarizes all of the calculations for applying the F test. The table includes the sum of squares between, the sum of squares within, the sum of squares total, the degrees of freedom between, the degrees of freedom within, the degrees of freedom total, the mean square between, the mean square within, and the F ratio.

Multiple Comparison Procedures: Statistical tests used to determine the nature of the relationship after the omnibus null hypothesis has been rejected for the overall F test.

Tukey HSD Test: A multiple comparison procedure used to compare all possible pairwise differences between population means.

Practice Questions: True-False Format

1. The one-way analysis of variance is typically used to analyze the relationship between two variables when the independent variable has three or more levels and is between subjects in nature and the dependent variable is quantitative and measured on a level that approximates interval characteristics.

2. The null hypothesis for one-way analysis of variance is that all of the population means defined by the levels of the independent variable are equal. *There is no relationship between*

3. In one-way analysis of variance, the alternative hypothesis states that a relationship exists such that the population means for the groups defined by levels of the independent variable are not all equal to one another.

4. The more similar the sample means for the levels defined by the independent variable are to one another, the more between-group variability there is.

5. Between-group variability in mean scores on the dependent variable reflects two things: (a) sampling error and (b) the effect of the independent variable on the dependent variable.

6. Within-group variability reflects sampling error and between-group treatment effects.

7. Greater variability of scores within a group is indicative of greater variability of scores within the corresponding population and, thus, a greater amount of sampling error.

8. The variance ratio can be represented as between-group variability divided by within-group variability.

9. When the null hypothesis is not true, we would expect the variance ratio to equal 1.00 most of the time.

10. When the null hypothesis is true, the between-group variability reflects only sampling error.

11. The sum of squares total reflects the total variability in the dependent variable across all individuals.

12. Another name given to the sum of squares between is the sum of squares error because it reflects the effects of disturbance variables and, hence, sampling error.

13. The total variability in scores can be expressed in terms of (1) the variability between the group means (between-group variability) and (2) the variability of deviations from the group means (within-group variability).

14. The variance ratio of between-group variability divided by within-group variability that is computed to test the null hypothesis does not utilize measures of mean squares, but rather the variance ratio is based on measures of sums of squares.

15. A mean square is simply a sum of squares divided by its corresponding standard error. $\frac{SS}{Df}$

16. The degrees of freedom associated with the sum of squares between is equal to N - k. $K-1$

17. As with the sum of squares in a one way ANOVA, the degrees of freedom are additive; if we sum the degrees of freedom between and the degrees of freedom within, we will obtain the degrees of freedom associated with the sum of squares total.

18. When sample sizes are equal, the mean square between is equivalent to the variance estimate of the sample means multiplied by n to reflect the number of scores on which each mean is based.

19. The terminology "mean square within" is merely another name for the standard deviation of scores within a level defined by the independent variable. $p. 330$ pooled variance estimate F

20. The variance ratio, formally referred to as the F ratio, is the mean square between divided by the mean square within.

21.The reason why measures of mean squares rather than measures of sums of squares are used to define the F ratio is that when relevant conditions are met, use of measures of mean squares yield a sampling distribution that closely approximates an F distribution, whereas measures of sums of squares do not.

22. The F distribution has the same shape regardless of the degrees of freedom associated with it.

23. In all F distributions, the most common value is 0.

24. Since all departures from the null hypothesis are reflected in the upper tail of the F distribution (as defined by the critical value), the F test is, by its nature, directional.

25. The F test is quite robust to violations of the normality and homogeneity of variance assumptions, particularly when the sample sizes are moderate (greater than 20) to large in size and the same for all groups.

26. The F distribution bears a mathematical relationship to the t distribution such that $F = t^2$.

27. In one-way analysis of variance, the strength of the relationship between the independent and dependent variables is indexed by eta-squared.

28. The Tukey HSD test discerns the nature of the relationship by testing a null hypothesis for each possible pair of population means defined by the levels of the independent variable.

29. The problem with testing each pairwise null hypothesis by conducting three independent groups t tests is that multiple t tests increase the probability of making a Type II error for at least one of the tests beyond the probability specified by the alpha level. *Type I*

30. The rule for choosing between two competing hypotheses for the HSD test is the following: If the critical difference exceeds the absolute value of the difference between sample means involved in the comparison, then reject the null hypothesis. *critical value*

Answers to True-False Items

1. T	11. T	21. T
2. F	12. F	22. F
3. T	13. T	23. F
4. F	14. F	24. F
5. T	15. F	25. T
6. F	16. F	26. T
7. T	17. T	27. T
8. T	18. T	28. T
9. F	19. F	29. F
10. T	20. T	30. F

Practice Questions: Short Answer

1. When is the one-way analysis of variance typically used to analyze the relationship between two variables?

2. How is the alternative hypothesis for a one-way analysis of variance different from previous tests?

3. What is between-group variability and what two factors contribute to it?

4. What is within-group variability and what contributes to it?

5. What is the F ratio?

6. Distinguish between the sum of squares between, the sum of squares within, and the sum of squares total.

7. How is the total variability in a set of scores partitioned in a one-way analysis of variance?

8. Why is the computation of the F ratio based on mean squares instead of sums of squares? Why?

9. What are the assumptions underlying the F test?

10. Describe the logic underlying the use of multiple comparison procedures.

11. What is the Tukey HSD test and why is it used instead of multiple independent groups t tests to discern the nature of the relationship?

Answers to Short Answer Questions

1. The one-way analysis of variance is typically used to analyze the relationship between two variables when:(1) the dependent variable is quantitative in nature and is measured on a level that at least approximates interval characteristics; (2) the independent variable is between-subjects in nature (it can be either qualitative or quantitative); and (3) the independent variable has three or more levels.

2. Unlike previous tests, the alternative hypothesis for a one-way analysis of variance cannot be summarized in a single mathematical statement. This is because there are many ways in which population means can pattern themselves so that they are not all equal to one another. Thus, the alternative hypothesis is: The three population means are not all equal.

3. Between-group variability reflects differences between the mean scores in the groups defined by the levels of the independent variable. If the mean scores in the three groups are all equal, then there is no between-group variability. The two factors that contribute to between-group variability are sampling error and the effect of the independent variable on the dependent variable.

4. Within-group variability reflects the variability of scores within each of the groups defined by the levels of the independent variable. Because the independent variable is held constant for each group, the variability in scores cannot be attributed to it. Other factors are operating to cause the variability in scores. Greater variability of scores within a group is indicative of

greater variability of scores within the corresponding population and, thus, a greater amount of sampling error.

5. The F ratio is the mean square between divided by the mean square within. It is used to test the viability of the null hypothesis in one way analysis of variance problems.

6. The sum of squares total reflects the total variability in the dependent variable across all individuals. The sum of squares between reflects between-group variability, or variability due to differences between the group means. The sum of squares within reflects within-group variability. Another name given to the sum of squares within is the sum of squares error because it reflects the effects of disturbance variables and, hence, sampling error.

7. In one-way analysis of variance, the total variability in the dependent variable can be partitioned into two components, such that

$$SS_{TOTAL} = SS_{BETWEEN} + SS_{WITHIN}$$

8. The reason why measures of mean squares rather than measures of sums of squares are used to define the F ratio is that when the relevant conditions are met, use of measures of mean squares yield an F ratio that has a sampling distribution that closely approximates an F distribution, whereas use of measures of sums of squares do not.

9. The sampling distribution of the F ratio tends to closely approximate an F distribution when the following assumptions are met: (1) the samples are independently and randomly selected from their respective populations; (2) the scores in each population are normally distributed; (3) the scores in each population have equal variances.

10. The F test considers the null hypothesis against all possible alternatives. If any one of the alternatives holds, the null hypothesis will be rejected, unless a Type I error occurs. However, the alternative hypothesis states only that the population means are not all equal. Given this state of affairs, it becomes necessary to conduct additional analyses to determine the exact nature of the relationship between the two variables when three or more groups are involved. Multiple comparison procedures do this.

11. The Tukey HSD test discerns the nature of the relationship by testing a null hypothesis for each possible pair of group means. We do not test each of these null hypotheses by conducting multiple independent groups t tests because multiple t tests increase the probability of making a Type I error for at least one of the tests beyond the probability specified by the alpha level. The Tukey HSD test circumvents this problem and maintains the probability of making at least one Type I error at the specified alpha level.

Answers to Selected Exercises from Textbook

Exercise Number 16: Exercise 16 provides the ANOVA summary table, and asks for the multiple comparisons by using the Tukey HSD procedure. It will be necessary to compute the critical difference in order to conduct the Tukey HSD. To see what we need to compute the HSD, write the formula for the critical difference (Equation 12.17).

$$CD = q \sqrt{\frac{MS_{WITHIN}}{n}}$$

The mean square within is given in the summary table as 5.00. The exercise also tells us that there are 10 subjects in each of 3 groups, so n is 10. The value of q comes from Appendix G. To use Appendix G we need the degrees of freedom error which, as cited in the summary table, is 27; the alpha level, which we will assume is .05; and the number of groups, which is 3. Since 27 degrees of freedom do not appear in the table, we use interpolation to obtain the value of q. Since 27 degrees of freedom are halfway between 24 df at 3.53 and 30 df at 3.49, we can add 3.53 and 3.49 to give 7.02, and divide by 2 to give 3.51. The value of q at 27 df is 3.51. Now we can substitute the values into the HSD formula:

$$CD = 3.51 \sqrt{\frac{5.00}{10}} = 2.48$$

The exercise provides the values of three group means. To finish, we only need to compare the various combinations of two group means to Tukey's critical difference (CD) computed above:

Null Hypothesis tested	Absolute difference between sample means	Value of CD	Null Hypothesis rejected?
$\mu_S = \mu_M$	$\lvert 6.00 - 8.00 \rvert = 2.00$	2.48	No
$\mu_S = \mu_D$	$\lvert 6.00 - 10.00 \rvert = 4.00$	2.48	Yes
$\mu_M = \mu_D$	$\lvert 8.00 - 10.00 \rvert = 2.00$	2.48	No

The nature of the relationship between marital status and attitudes toward divorce is that divorced individuals ($\overline{X}_D = 10.00$) have more positive attitudes than single people ($\overline{X}_D = 6.00$). However, we cannot confidently conclude that either divorced or single individuals differ from married individuals ($\overline{X}_D = 8.00$) in their attitudes toward divorce.

Exercise 46: In this exercise, the independent variable is the defendant's race, with three levels: White, African American and Hispanic. The dependent variable is the judgment of the probability of the defendant's guilt. The independent variable is between-subjects and has more than two levels. The dependent variable is quantitative and measured on approximately an interval level. Consequently, we use the one-way analysis of variance. Exercise 46 tells us to analyze the data and draw a conclusion. After making sure you understand the nature of the study, and selecting the appropriate test, begin by stating the hypotheses. The null hypothesis is that the defendant's race does not influence judgments of the probability of guilt. The alternative hypothesis is that the defendant's race does influence judgments of the probability of guilt:

$$H_0: \mu_W = \mu_A = \mu_H$$

H_1: The three population means are not all equal.

We will do the computations necessary for the three sums of squares first, and then substitute the appropriate values into each formula:

White Defendant		AA Defendant		Hispanic Defendant	
X	X²	X	X²	X	X²
6	36	10	100	10	100
7	49	10	100	6	36
2	4	9	81	10	100
3	9	4	16	5	25
5	25	4	16	10	100
0	0	10	100	5	25
1	1	10	100	2	4
0	0	10	100	10	100
6	36	3	9	2	4
0	0	10	100	10	100

138

$T_W = 30$ $\qquad\qquad$ $T_A = 80$ $\qquad\qquad$ $T_H = 70$

$\overline{X}_W = 3.00$ $\qquad\qquad$ $\overline{X}_A = 8.00$ $\qquad\qquad$ $\overline{X}_H = 7.00$

$T_W^2 = 900$ $\qquad\qquad$ $T_A^2 = 6,400$ $\qquad\qquad$ $T_H^2 = 4,900$

$\Sigma X^2 = 36 + 49 + \ldots + 4 + 100 = 1,476$

$\Sigma X = 6 + 7 + \ldots + 2 + 10 = 180$

$$\frac{(\Sigma X)^2}{N} = \frac{(180)^2}{30} = 1,080.00$$

$$\frac{\Sigma T_j^2}{n} = \frac{900 + 6,400 + 4,900}{10} = 1,220.00$$

Now, substitute the appropriate values into the sum of squares formulas (Equations 12.13, 12.12, 12.11).

$$SS_{BETWEEN} = \frac{\Sigma T_j^2}{n} - \frac{(\Sigma X)^2}{N}$$

$$= 1,200.00 - 1,080.00 = 140.00$$

$$SS_{WITHIN} = \Sigma X^2 - \frac{\Sigma T_j^2}{n}$$

$$= 1{,}476 - 1{,}220 = 256.00$$

$$SS_{TOTAL} = \Sigma X^2 - \frac{(\Sigma X)^2}{N}$$

$$= 1{,}476 - 1{,}080 = 396.00$$

Next, we compute the degrees of freedom (Equations 12.6, 12.7, 12.9).

$$df_{BETWEEN} = k - 1 = 3 - 1 = 2$$

$$df_{WITHIN} = N - k = 30 - 3 = 27$$

$$df_{TOTAL} = N - 1 = 30 - 1 = 29$$

Now, the mean squares can be computed (Equations 12.4, 12.5).

$$MS_{BETWEEN} = \frac{SS_{BETWEEN}}{df_{BETWEEN}} = \frac{140.00}{2} = 70.00$$

$$MS_{WITHIN} = \frac{SS_{WITHIN}}{df_{WITHIN}} = \frac{256.00}{27} = 9.48$$

Finally, we can compute the F ratio (Equation 12.10).

$$F = \frac{MS_{BETWEEN}}{MS_{WITHIN}} = \frac{70.00}{9.48} = 7.38$$

The calculations yield the following source table:

Source	SS	df	MS	F
Between	140.00	2	70.00	7.38
Within	356.00	27	9.48	
Total	396.00	29		

We are ready to test the observed F for significance. For 2 degrees of freedom between, 27 degrees of freedom within, and an alpha level of .05, the critical values in Appendix F are ±3.35. The observed value of F is 7.38, which exceeds the positive critical value of 3.35. Therefore, we reject the null hypothesis and conclude that a relationship exists between the independent variable (the defendant's race) and the dependent variable (the judgments of the probability of guilt).

Next, we evaluate the strength of the relationship with eta-squared (Equation 12.15).

$$eta^2 = \frac{SS_{BETWEEN}}{SS_{TOTAL}} = \frac{140.00}{396.00} = .35$$

The eta-squared of .35 represents a strong effect, and indicates that 35% of the variability in judgments of the probability of guilt is due to the defendant's race.

Finally, we test the nature of the relationship with Tukey's HSD (Equation 12.17).

$$CD = q\sqrt{\frac{MS_{WITHIN}}{n}}$$

The mean square within is reported in the summary table as 9.48. The number of subjects in each group is 10. From Appendix G, the value of q for an alpha of .05, with 27 degrees of freedom within, and 3 groups is 3.51 (by interpolation).

$$CD = 3.51 \sqrt{\frac{9.48}{10}} = 3.42$$

Null Hypothesis tested	Absolute difference between sample means	Value of CD	Null Hypothesis rejected?
$\mu_W = \mu_A$	$\mid 3.00 - 8.00 \mid = 5.00$	3.42	Yes
$\mu_W = \mu_H$	$\mid 3.00 - 7.00 \mid = 4.00$	3.42	Yes
$\mu_A = \mu_H$	$\mid 8.00 - 7.00 \mid = 1.00$	3.42	No

The nature of the relationship is such that judgments of the probability of guilt are more likely when the defendant is African American ($\overline{X}_A = 8.00$) than when the defendant is white ($\overline{X}_W = 8.00$). Judgments of the probability of guilt are also more likely when the defendant is Hispanic ($\overline{X}_M = 8.00$) than when the defendant is white. We cannot confidently conclude that judgments of the probability of guilt are more likely with African Americans compared to Hispanics.

Exercise 46 asks us to write the results using the principles discussed in the Method of Presentation section (Note: All standard deviations were computed using Equation 7.2 for a standard deviation estimate in Chapter 7):

Results

A one-way analysis of variance was conducted on judgments of the probability of guilt as a function of the defendant's race (white, African American, or Hispanic), using an alpha level of .05. The F was statistically significant ($\underline{F}(2, 27) = 7.38$, $\underline{p} < .01$). The strength of the relationship, as indexed by eta-squared, was .35. A Tukey's HSD test revealed that the mean guilt-probability judgment for the white defendant (M = 3.00, SD = 2.79) was significantly lower than the mean guilt-probability judgment for either the African American (M = 8.00, SD = 3.02) or the Hispanic (M = 7.00, SD = 3.40) defendant. The mean guilt-probability judgments for the African American and the Hispanic defendants did not significantly differ.

Chapter 13: One-Way Repeated Measures Analysis of Variance

Study Objectives

This chapter presents one-way repeated measures analysis of variance, a test that examines the relationship between two variables. After reading the material in this chapter, you should be able to specify the conditions when you would use one-way repeated measures analysis of variance to analyze a bivariate relationship. You should be able to define and characterize between group variability, variability across subjects and error variability and the statistical measures of them: The sum of squares IV, the mean square IV, the sum of squares error, and the mean square error, the sum of squares across subjects and the mean square across subjects. You should be able to describe the logic of the F test and the F ratio. You should be able to apply one-way repeated measures analysis of variance in its entirety to determine if a relationship exists between two variables. This includes knowing how to compute all of the intermediate statistics and forming a summary table for the calculations. You should be able to specify the assumptions underlying one-way repeated measures analysis of variance and briefly characterize the robustness of the test to assumption violations. You should be aware of the Huyhn-Feldt adjustment procedure in cases where sphericity assumptions are violated.

You should be able to calculate and interpret eta squared for one-way repeated measures analysis of variance. You should be able to determine the nature of a relationship, using the Tukey HSD test. You should be able to explain why the Tukey HSD test is used instead of multiple t tests and you should be aware of testing strategies for pairwise contrasts when the assumption of sphericity is violated.

You should be able to determine the sample size necessary to achieve a given level (e.g., 0.80) of statistical power when designing a study that will use one-way repeated measures analysis of variance. You should be able to explain how alpha, sample size, and the value of eta squared in the population influences statistical power for this test.

You should be able to write up the results of a one-way repeated measures analysis of variance using APA format as discussed in the Method of Presentation section. You should be able to interpret the results of a repeated measures analysis of variance from examples reported in the literature.

Study Tips

The computational formulas in this chapter, although straight-forward, are somewhat complicated and students tend to make more computational errors in them than in other chapters throughout the book. Make sure to double check your calculations with a calculator.

Some students have a difficult time understanding the concept of across subject variability and how this reflects individual backgrounds. Across-subject variability is based on a person's average score across experimental conditions, with the idea being that people with higher average scores differ in their backgrounds from people with lower average scores. Suppose you and a friend are asked to rate each of five wines on a 1 to 10 scale (with higher scores indicating a better taste) and your average rating is 8 and your friend's average rating is 2. It would seem logical to conclude that you like wine better than your friend does. This may be because your friend has not had much experience with wine (at first, wine tastes sour, but then some people acquire a taste for it), or s/he got very sick on it at sometime in the past, or s/he doesn't like alcohol in general, or any other of a host of reasons pertaining to your different backgrounds and experiences. These individual differences *are* reflected in your average ratings and are statistically controlled for in a repeated measures analysis of variance.

Glossary Of Important Terms

Study the terms listed below. Make sure you understand each so that you could explain them to someone else who does not know them.

One-Way Repeated Measures Analysis of Variance: A statistical technique that is used to analyze the relationship between two variables, under the same circumstances as the correlated groups t test except that the independent variable has more than two levels. The procedure is focused on testing the null hypothesis of equivalent population means across the levels of the independent variable.

Sum of Squares IV: A sum of squares that reflects between group variability in a repeated measures design. It is based on the mean scores in each condition and the sample size.

Sum of Squares Across Subjects: A sum of squares based on each research participant's average (mean) score across conditions. It reflects variability due to individual backgrounds.

Sum of Squares Error: A sum of squares as applied to deviation scores across all individuals in the study, after the effects of individual background have been removed from the data. The deviation scores are the raw score minus the mean of the condition in which the score occurs. It reflects within-group variability (after removing the effects of individual background).

Mauchly Test: A statistical test that evaluates the potential violation of the sphericity assumption in a repeated measures design.

Adjustment Factors: Adjustments to the traditional F test when the assumption of sphericity is violated. The adjustments are to the degrees of freedom associated with the F ratio. The two most common are the Huyhn-Feldt and Geisser-Greenhouse adjustment procedures.

Modified Bonferroni Procedure: A procedure used in conjunction with the correlated groups t test for performing all possible pairwise comparisons in order to discern the nature of a relationship after a statistically significant omnibus F test in a repeated measures analysis of variance.

Practice Questions: True-False Format

1. The one-way repeated measures analysis of variance is typically used to analyze the relationship between two variables when the independent variable has two and only two levels.

2. The one-way repeated measures analysis of variance is used under the same conditions as one-way between-subjects analysis of variance except that the independent variable is within-subjects in nature rather than between-subjects in nature.

3. The one-way repeated measures analysis of variance is an extension of the independent groups t test.

4. Nullified scores cannot be generated when a within-subjects independent variable has more than two levels.

5. In a one-way repeated measures analysis of variance, the total variability in the dependent variable across all individuals and all conditions can be represented by the sum of squares total.

6. In a one-way repeated measures analysis of variance, the total variability in the dependent variable can be partitioned into three components, one reflecting the influence of the independent variable (called the sum of squares IV), one reflecting the influence of

individual differences (called the sum of squares across subjects), and one reflecting the influence of disturbance variables other than individual differences (called the sum of squares error).

7. The sum of squares IV is conceptually equivalent to the "sum of squares within groups" in the one-way between-subjects analysis of variance.

8. The sum of squares error is conceptually equivalent to the "sum of squares within groups" in the one-way between-subjects analysis of variance.

9. The sum of squares across subjects has no counterpart in the one-way between-subjects analysis of variance.

10 When a within-subjects design is used, each score is derived from a different person, so it is not possible to estimate the effects of disturbance variables due to individual differences.

11. The sum of squares across subjects is a feature that differentiates the between-subjects and the repeated measures analysis of variance.

12. When a repeated measures design is used, the influence of individual differences cannot be estimated or statistically removed from the dependent variable.

13. The sum of squares within in a between-subjects design does not include the effects of individual differences while the sum of squares error in a repeated measures design does.

14. The mean square error, which forms the denominator in the F test of the relationship between the independent and dependent variables in the repeated measures case, will tend to be smaller than the mean square within forming the denominator of the F test in the between-subjects case.

15. To the extent that the mean square within is larger than the mean square error, a larger F ratio and a greater likelihood of rejecting the null hypothesis should result in a between-subjects analysis of variance as opposed to a repeated measures analysis of variance.

16. In a repeated measures design, the sum of squares error represents the influence of disturbance variables other than individual differences--that is, variability in the dependent variable remaining after the effects of the independent variable and individual differences have been partitioned out.

17. As with the sum of squares in a repeated measures ANOVA, the degrees of freedom are additive, such that $df_{TOTAL} = df_{IV} + df_{ACROSS\ SUBJECTS} + df_{ERROR}$.

18. In order to test the null hypothesis of equivalent population means, it is necessary to calculate mean squares for the IV, error, and across subjects components.

19. The mean square across subjects does not directly figure in the F test constituting the hypothesis testing procedure and, thus, is not typically calculated.

20. The main function of the sum of squares across subjects is to remove variability due to individual differences from the independent variable so that a more sensitive test of the relationship between the independent and dependent variables can be performed.

21. The F ratio for a one-way repeated measures analysis of variance is MS_{IV} divided by $MS_{SUBJECTS}$.

22. When reporting an F ratio, it is conventional to present the degrees of freedom IV followed by the degrees of freedom error.

23. The F test for one-way repeated measures analysis of variance is appropriate when the dependent variable is qualitative in nature.

24. The usual assumption that the sample is independently and randomly selected from the population of interest is not applicable to the repeated measures analysis of variance.

25. The validity of the F test for one-way repeated measures analysis of variance is based on the assumption that the variance of the population difference scores for any two conditions is the same as the variance of the population difference scores for any other two conditions.

26. The assumption that the variance of the population difference scores for any two conditions is the same as the variance of the population difference scores for any other two conditions is known as the non-homogeneous differences assumption.

27. Although the F test is robust to violations of the sphericity assumption, it is not robust to violations of the normality assumption.

28. Among the alternatives to the traditional F test when the sphericity assumption is violated, two of the most frequently encountered adjustment factors are the Huyhn-Feldt epsilon and the Greenhouse-Geisser epsilon.

29. As with the correlated groups t test, eta-squared in the context of the one-way repeated measures analysis of variance represents the proportion of variability in the dependent variable that is associated with the independent variable, before variability due to individual differences has been removed.

30. If the assumption of sphericity is satisfied, the Tukey HSD test cannot be used to discern the nature of the relationship in a one-way repeated measures analysis of variance.

31. Since the value of F required to reject the null hypothesis becomes more extreme as the degrees of freedom become smaller, repeated measures analysis of variance might actually be less powerful than between-subjects analysis of variance when individual differences have only a minimal influence on the dependent variable.

Answers to True-False Items

1. F	11. T	21. F	31. T
2. T	12. F	22. T	
3. F	13. F	23. F	
4. F	14. T	24. F	
5. T	15. F	25. T	
6. T	16. T	26. F	
7. F	17. T	27. F	
8. T	18. F	28. T	
9. T	19. T	29. F	
10. F	20. F	30. F	

Practice Questions: Short Answer

1. When is the one-way repeated measures analysis of variance typically used to analyze the relationship between two variables?

2. How is the total variability in the dependent variable partitioned in a repeated measures design?

3. Compare the different sums of squares for a repeated measures design with the sums of squares for a between-subjects design.

4. Why is a repeated measures analysis of variance usually a more sensitive test of the relationship between the independent and dependent variables than is a between-subjects analysis of variance?

5. What does the mean square subjects contribute to the F test?

6. What are the assumptions underlying the F test for a one-way repeated measures analysis of variance?

7. Discuss the issue of the robustness of the F test to violations of the sphericity assumption.

8. How is the nature of the relationship addressed in a one-way repeated measures analysis of variance?

9. Under what circumstances might repeated measures analysis of variance actually be less powerful than between-subjects analysis of variance?

10. Describe an alternative to counterbalancing as a means of controlling confounding variables.

Answers to Short Answer Questions

1. The one-way repeated measures analysis of variance is typically used to analyze the relationship between two variables when:(1) the dependent variable is quantitative in nature and is measured on a level that at least approximates interval characteristics; (2) the independent variable is within-subjects in nature (it can be either qualitative or quantitative); and (3) the independent variable has three or more levels.

2. Like the between-subjects design, the total variability in the dependent variable in a repeated measures design across all individuals and all conditions can be represented by the sum of squares total. This can be partitioned into three components, one reflecting the independent variable (called the sum of squares IV), one reflecting the influence of individual differences (called the sum of squares across subjects), and one reflecting the influence of disturbance variables other than individual differences (called the sum of squares error). Symbolically,

$$SS_{TOTAL} = SS_{IV} + SS_{ACROSS\ SUBJECTS} + SS_{ERROR}$$

3. The sum of squares IV is conceptually equivalent to the "sum of squares between groups" in the between-subjects design. The sum of squares error and the "sum of squares within" are also conceptually equivalent as both reflect only the influence of disturbance variables. The sum of squares across subjects has no counterpart in the between-subjects design. It reflects the influence of individual differences only.

4. When a between-subjects design is used, each score is derived from a different person, so it is not possible to estimate the effects of disturbance variables due to individual differences. Differences in background thus contribute to sampling error, as reflected in the sum of squares within. However, when a repeated measures design is used, the influence of individual differences can be estimated and statistically removed from the dependent variable. The sum of squares within in a between-subjects design includes the effects of individual differences while the sum of squares error in a repeated measures design does not. Thus, the mean square error, which forms the denominator in the F test of a relationship between the independent and dependent variables in the repeated measures case, will tend to be smaller than the mean square within forming the denominator of the F test in the between-subjects case.

5. The mean square across subjects does not directly figure in the F test constituting the hypothesis testing procedure and, thus, is not typically calculated.

6. The F test for one-way repeated measures analysis of variance is appropriate when the dependent variable is quantitative in nature and measured on a level that at least approximates interval level characteristics. Its validity rests on the following assumptions: (1) the sample is independently and randomly selected from the population of interest; (2) each population of scores is normally distributed; (3) the variance of the population difference scores for any two conditions is the same as the variance of the population difference scores for any other two conditions.

8. The nature of the relationship following a statistically significant one-way repeated measures analysis of variance is addressed differently, depending on whether the assumption of sphericity is satisfied. If the assumption is satisfied, then one applies the Tukey HSD test. If the sphericity assumption is not met, then the HSD test is not appropriate for evaluating the nature of the relationship. In this case, we use a modified Bonferroni procedure coupled with a set of correlated groups t tests.

9. The degrees of freedom for the denominator of the F test will always be less in the repeated measures case than in the between-subjects case. Since the value of F required to reject the null hypothesis becomes more extreme as the degrees of freedom become smaller, repeated measures analysis of variance might actually be less powerful than between-

subjects analysis of variance when individual differences have only a minimal influence on the dependent variable.

10. An alternative to counterbalancing as a means of controlling confounding variables is to randomly order the conditions for each subject. The rationale is that when the sequence of conditions across subjects is randomly determined, chance will ensure that each condition occurs in each position an approximately equal number of times and, thus, that carry-over effects are evenly distributed across conditions. This approach is particularly valuable when the number of conditions is so great that counterbalancing is impossible.

Answers to Selected Exercises from Textbook

Exercise Number 44: In this exercise, the independent variable is the factors about which males would be most concerned when evaluating male oral contraceptives. The independent variable has four levels: health risks, effectiveness, cost, and convenience. The dependent variable is the rating of importance of each factor, measured using a 21-point scale, with higher numbers indicating higher degrees of importance.

The independent variable is within-subjects and has three or more levels. The dependent variable is quantitative and measured on approximately an interval level. Therefore, a one-way repeated measures analysis of variance is appropriate. Let us lay out the calculations:

	Health Risks		Effectiveness		Cost		Convenience			
Individual	X	X^2	X	X^2	X	X^2	X	X^2	s_i	s_i^2
1	20	400	12	144	8	64	8	64	48	2,304
2	19	361	15	225	11	121	11	121	56	3,136
3	18	324	14	196	10	100	10	100	52	2,704
4	17	289	13	169	9	81	9	81	48	2,304
5	16	256	16	256	12	144	12	144	56	3,136

$T_H = 90$ $T_E = 70$ $T_C = 50$ $T_{CV} = 50$

$\overline{X}_H = 18.00$ $\overline{X}_E = 14.00$ $\overline{X}_C = 10.00$ $\overline{X}_{CV} = 10.00$

$T_H^2 = 8,100$ $T_E^2 = 4,900$ $T_C^2 = 2,500$ $T_{CV}^2 = 4,900$

$\Sigma X^2 = 400 + 361 + ... + 81 + 144 = 3,640$

$\Sigma X = 20 + 19 + ...+ 9 + 12 = 260$

$$\frac{(\Sigma X)^2}{k N} - \frac{(260)^2}{(4)(5)} = 3,380.00$$

$$\frac{\Sigma T_j^2}{N} = \frac{8,100 + 4,900 + 2,500 + 2,500}{5} = 3,600.00$$

$$\frac{\Sigma s_i^2}{k} = \frac{2,304 + 3,136 + 2,704 + 2,304 + 3.136}{4} = 3,396.00$$

Now we can substitute the appropriate values into the computational formulas for the sums of squares (Equations 13.2, 13.3, 13.4, 13.5).

$$SS_{TOTAL} = \Sigma X^2 - \frac{(\Sigma X)^2}{kN}$$

$$= 3,640 - 3,380 = 260.00$$

$$SS_{IV} = \frac{\Sigma T_j^2}{N} - \frac{(\Sigma X)^2}{kN}$$

$$= 3,600.00 - 3,380.00 = 220.00$$

$$SS_{\text{ACROSS SUBJECTS}} = \frac{\Sigma s_i^2}{k} - \frac{(\Sigma X)^2}{kN}$$

$$= 3{,}396.00 - 3{,}380.00 = 16.00$$

$$SS_{\text{ERROR}} = \Sigma X^2 + \frac{(\Sigma X)^2}{kN} - \frac{\Sigma T_j^2}{N} - \frac{\Sigma s_i^2}{k}$$

$$= 3{,}640 + 3{,}380 - 3{,}600 - 3{,}396 = 24.00$$

We are ready to compute the degrees of freedom associated with each sum of squares (Equations 12.6, 12.7, 12.9):

$$df_{\text{IV}} = k - 1 = 4 - 1 = 3$$

$$df_{\text{ACROSS SUBJECTS}} = N - 1 = 5 - 1 = 4$$

$$df_{\text{ERROR}} = (k - 1)(N - 1) = (4 - 1)(5 - 1) = 12$$

$$df_{\text{TOTAL}} = kN - 1 = (4)(5) - 1 = 19$$

Next, we need the mean squares (Equations 13.11, 13.12).

$$MS_{\text{IV}} = \frac{SS_{\text{IV}}}{df_{\text{IV}}} = \frac{220.00}{3} = 73.33$$

$$MS_{\text{ERROR}} = \frac{SS_{\text{ERROR}}}{df_{\text{ERROR}}} = \frac{24.00}{12} = 2.00$$

The F ratio for the one-way repeated measures analysis of variance is next (Equation 13.13).

$$F = \frac{MS_{IV}}{MS_{ERROR}} = \frac{73.33}{2.00} = 36.66$$

We can now summarize our computations in a summary table:

Source	SS	df	MS	F
IV	220.00	3	73.33	36.66
Error	24.00	12	2.00	
Across Subjects	16.00	4	9.48	
Total	260.00	19		

The critical values of F from Appendix F for an alpha of .05 with 3 and 12 degrees of freedom are ±3.49. The observed F of 36.66 is greater than the positive critical value of 3.49, so we reject the null hypothesis and conclude that a relationship exists between the four factors and the ratings of the importance of the factors.

We use eta-squared to evaluate the strength of the relationship (Equation 13.14):

$$eta^2 = \frac{SS_{IV}}{SS_{IV} + SS_{ERROR}}$$

The two sums of squares we need are in the ANOVA summary table.

$$eta^2 = \frac{220.00}{220.00 + 24.00} = .90$$

The proportion of variability in the ratings of the four factors that is associated with the importance of the factors after the effects of individual differences have been removed is .90. This represents a strong relationship between the independent and dependent variables.

We use Tukey's HSD to test the nature of the relationship (Equation 13.16).

$$CD = q \sqrt{\frac{MS_{WITHIN}}{n}}$$

We need to find the value of q from Appendix G. For an overall alpha level of .05, degrees of freedom error of 12, and k of 4, the Appendix G value of q is 4.20. We can pick up the mean square error from the ANOVA summary table, and N is 5.

$$CD = 4.20 \sqrt{\frac{2.00}{5}} = 2.66$$

The critical difference is thus 2.66. Now we can finish up the HSD analysis by comparing the difference between all possible combinations of two group means to the CD value.

Null Hypothesis tested	Absolute difference between sample means	Value of CD	Null Hypothesis rejected?
$\mu_H = \mu_E$	$\mid 18.00 - 14.00 \mid = 4.00$	2.66	Yes
$\mu_H = \mu_C$	$\mid 18.00 - 10.00 \mid = 8.00$	2.66	Yes
$\mu_H = \mu_{CV}$	$\mid 18.00 - 10.00 \mid = 8.00$	2.66	Yes
$\mu_E = \mu_C$	$\mid 14.00 - 10.00 \mid = 4.00$	2.66	Yes
$\mu_E = \mu_{CV}$	$\mid 14.00 - 10.00 \mid = 4.00$	2.66	Yes
$\mu_C = \mu_{CV}$	$\mid 10.00 - 10.00 \mid = 0.00$	2.66	No

The nature of the relationship is such that health risks are rated higher than effectiveness, cost, and convenience; effectiveness is rated higher than costs and convenience; and cost and convenience are rated the same.

Exercise 44 also asks us to write the results using the principles presented in the Method of Presentation section (Note: All standard deviations were computed using Equation 7.2 for a standard deviation estimate in Chapter 7):

Results

A one-way repeated measures analysis of variance compared the mean ratings of how important four factors (health risks, effectiveness, cost, and convenience) were considered to be in evaluating male oral contraceptives, using an alpha level of .05. The observed F ratio was statistically significant, $F(3, 12) = 36.66$, $p < .01$. The strength of the relationship, as indexed by eta-squared, was .90. A Tukey's HSD test indicated that the health risks factor ($M = 14.00$, $SD = 1.58$) was rated as statistically significantly more important than either the cost ($M = 10.00$, $SD = 1.58$) or the convenience ($M = 10.00$, $SD = 1.58$) factor, but that none of the other pairwise contrasts were statistically significant.

Chapter 14: Pearson Correlation and Regression: Inferential Aspects

Study Objectives

This chapter covers inferential aspects of correlation and regression. After reading the material, you should be able to specify the conditions when you would use Pearson correlation. You should be able to explain the sampling distribution of a correlation coefficient and characterize its basic shape. You should understand the difference between a population regression equation and a sample regression equation. You should be able to explain and characterize the residual term (ϵ) in a regression equation. You should be able to compute a Pearson correlation from sample data and answer all three fundamental questions about it (is there a relationship between two variables, what is the strength of the relationship, what is the nature of the relationship). You should be able to calculate the t statistic for testing if a non-zero population correlation exists and you should be able to use the tabled values of r in the Appendix to perform a test of the null hypothesis.

You should be able to identify situations where you would be interested in calculating a regression equation. You should be able to compute and interpret an intercept, a slope, and an estimated standard error of estimate.

You should be able to state the assumptions underlying the significance tests of Pearson correlation and how robust the technique is to assumption violations.

You should be able to determine the sample size necessary to achieve a given level (e.g., 0.80) of statistical power when designing a study that will use Pearson's correlation coefficient. You should be able to explain how alpha, sample size, and the value of the squared correlation in the population influences statistical power for this test.

You should be able to write up the results of a Pearson correlation analysis using APA format as discussed in the Method of Presentation section. You should be able to interpret the results of a correlation analysis from examples reported in the literature.

Study Tips

Most of the material in this chapter is an extension of the material in Chapter 5. Make sure you re-read and carefully study Chapter 5 before reading this chapter.

A crucial point to keep in mind is the distinction between a population regression equation and a sample regression equation. Students often fail to maintain the proper distinction, and use the two interchangeably. For two variables, X and Y, it is possible, in principle to calculate a regression equation in the *population*. In this case, the regression equation/ linear model has the form

$$Y = \alpha + \beta X + \epsilon$$

where α is the least squares value for the intercept, β is the least squares value for the slope, and ϵ is the residual term reflecting how far a given individual deviates from the population regression line. The values of α and β would be calculated using the equations in Chapter 5 *as applied to the entire population of scores.* Note the use of Greek notation, to indicate that we are dealing with population parameters.

If we select a random sample of cases from the population, then we can calculate the regression equation/ linear model for the sample, and it appears as follows:

$$Y = a + b X + e$$

where a is the least squares value for the sample intercept, b is the least squares value for the sample slope, and e is how far a given individual deviates from the sample regression line. Note the absence of Greek notation and how the equation has the same basic form as the population equation. We are typically interested in knowing the values of α and β as well as the standard deviation of the ϵ scores (which reflects the population standard error of estimate). However, we cannot know these values unless we collect data from everyone in the population. This is usually impossible. So we resort to selecting a random sample and using the sample intercept, a, as an estimate of α, and the sample slope, b, as an estimate of β. We estimate the standard deviation of ϵ using the estimated standard error of estimate, as described in the text.

Be sure to keep these distinctions in mind. They are crucial and students often confuse them.

Glossary Of Important Terms

Study the terms listed below. Make sure you understand each so that you could explain them to someone else who does not know them.

Error Score: A variable in the linear model that reflects all factors that are uncorrelated with X that influence Y.

Sampling Distribution of the Correlation Coefficient: A distribution of correlation coefficients based on all possible random samples of a certain size.

Bivariate Normal Distribution: A theoretical distribution of X and Y scores considered simultaneously that has known mathematical properties. For example, in a bivariate normal distribution, the distribution of Y scores at any value of X is normal.

Coefficient of Determination: The proportion of variability that is common to two variables. It is the same as r^2 or eta squared.

Criterion Variable: The variable that is being predicted (Y).

Predictor Variable: The variable from which predictions are made (X).

Estimated Standard Error of Estimate: An estimate of the population standard error of estimate based on sample data. It reflects an estimate of the average amount of error when predicting Y from X in the linear model.

Practice Questions: True-False Format

1. Pearson correlation is typically used to analyze the relationship between two variables when the observations on each variable are within-subjects in nature.

2. In practice, the most common use of correlation procedures involves making inferences about correlation coefficients in populations based on sample data.

3. The linear model, as applied to regression problems, has the form $Y = \alpha + \beta X + \epsilon$.

4. The linear model states that a person's score on Y is a linear function of X, with β representing the intercept and α representing the slope.

5. In the equation for the linear model, ϵ is called an error score that reflects all factors that are uncorrelated with X that influence Y.

6. Perfect linear relationships between Y and X are relatively common in the behavioral sciences.

7. If everyone in the population has an ϵ score of zero, then the variance of the ϵ scores across individuals is also zero, and Y is a perfect linear function of X.

160

8. If a relationship is observed between two variables in a set of sample data, this means that a relationship exists between the variables in the corresponding population.

9. A relationship between two variables might exist in a sample even though it does not exist in the population.

10. The mean of a sampling distribution of the correlation coefficient is approximately ρ, the true population coefficient.

11. When the population correlation coefficient is zero and the scores in the population are (bivariate) normally distributed, then as N increases the distribution of the sample correlation coefficients tends, somewhat slowly, toward a binomial distribution.

12. When $\rho \neq 0$, the sampling distribution of the correlation coefficient is skewed.

13. We can test the null hypothesis that $\rho = 0$ by transforming the sample correlation coefficient into a statistic that has a sampling distribution that closely approximates the F distribution with k - 1 degrees of freedom.

14. Because the calculation of t depends only on the value of r, it is impossible to determine values of r that will lead to a rejection of the null hypothesis that $\rho = 0$, given a certain sample size.

15. The test of the null hypothesis that $\rho = 0$ is based on the assumption that the variability of Y scores in the population is the same at each value of X.

16. The assumption that the population distributions of X and Y are such that their joint distribution (that is, their scatterplot) represents a bivariate normal distribution implies that the distribution of Y scores at any value of X is normal in the population.

17. Under certain conditions, the t test for the Pearson correlation is robust with respect to violations of normality and homogeneity of variance.

18. Variance homogeneity can be problematic for the t test of the correlation coefficient, especially if one is going to pursue estimation of the population regression equation.

19. r^2 is formally known as the coefficient of indetermination.

20. r^2 represent the proportion of variability that is shared by the two variables.

21. Another useful index of the strength of the effect of one variable on another is the intercept.

22. The slope indicates the number of units Y is predicted to change given a one unit change in X.

23. The sample statistic, b, is a biased estimate of β.

24. The use of the slope as an index of effect size only makes sense when the correlation between Y and X is relatively low.

25. The nature of the relationship between two correlated variables is determined through examination of the sign of the correlation coefficient observed in the sample.

26. A regression equation can be used to identify the value of Y that is predicted to be paired with an individual's score on X.

27. An important weakness of regression is that prediction procedures cannot be applied to individuals who were not included in the original data set.

28. In the context of regression, prediction merely refers to the fact that we are making inferences about one variable from a second variable and does not imply that the latter variable causes the former.

29. The variable being predicted, Y, is formally known as the dependent or criterion variable.

30. The variable from which predictions are made, X, is formally known as the independent or predictor variable.

31. The estimated standard error of estimate estimates the average error that will be made across individuals when predicting scores on X from the regression equation.

32. If X helps to predict Y, then the estimated standard error of estimate will be larger than the estimated standard deviation of Y, and the better the predictor X is, the larger the estimated standard error of estimate will be.

Answers to True-False Items

1. F	11. F	21. F	31. F
2. T	12. T	22. T	32. F
3. T	13. F	23. F	
4. F	14. F	24. F	
5. T	15. T	25. T	
6. F	16. T	26. T	
7. T	17. T	27. F	
8. F	18. F	28. T	
9. T	19. F	29. T	
10. T	20. T	30. T	

Practice Questions: Short Answer

1. When is Pearson correlation typically used to analyze the relationship between two variables?

2. In the linear model $Y = \alpha + \beta X + \epsilon$, what does the ϵ term represent?

3. What is a sampling distribution of the correlation coefficient?

4. What are the assumptions of Pearson correlation?

5. Discuss the robustness of Pearson correlation.

6. How is the strength of the relationship between two variables evaluated in a correlational analysis?

7. How is the nature of the relationship between two variables evaluated in a correlational analysis?

8. How can regression analysis be applied to individuals who were not included in the original data set?

9. What is the estimated standard error of estimate?

Answers to Short Answer Questions

1. The Pearson correlation is typically used to analyze the relationship between two variables when: (1) both variables are quantitative in nature and are measured on a level that at least approximates interval characteristics; (2) the two variables have been measured on the same individuals, and (3) the observations on each variable are between-subjects in nature.

2. It is rare in the behavioral sciences that a perfect linear relationship exists between Y and X and this is why the ϵ term is included in the equation for the linear model. This term is called an error score and reflects all factors that are uncorrelated with X that influence Y.

3. A sampling distribution of the correlation coefficient is a distribution of correlation coefficients based on all possible random samples of a given size.

4. The test of the null hypothesis that $\rho = 0$ is based on the following assumptions: (1) The sample is independently and randomly selected from the population of interest; (2) the population distributions of X and Y are such that their joint distribution represents a bivariate normal distribution; (3) the variability of Y scores is the same at each value of X.

5. While it is important that the assumption of independent and random selection of the sample from the population of interest not be violated, under certain conditions the t test for the Pearson correlation is robust with respect to violations of normality and homogeneity of variance. Research has shown that the t test of the correlation coefficient is robust to highly non-normal distributions when sample sizes are greater than 15. Variance heterogeneity can be problematic, especially if one is going to pursue estimation of the population regression equation.

6. The strength of the relationship between two variables in a correlational analysis can be represented r^2, which is the proportion of variance in Y associated with X, based on the linear model. Another potential index is the slope, but this makes sense only when the correlation between Y and X is relatively high.

7. The nature of the relationship between two variables is determined through examination of the sign of the correlation coefficient observed in the sample. If correlation coefficient is positive, the conclusion is that the population correlation coefficient is also positive and the variables approximate a direct linear relationship. If the sample correlation coefficient is negative, the appropriate conclusion is that the two variables approximate an inverse linear relationship in the population.

8. An important characteristic of regression is that prediction procedures can be extended to individuals who were not included in the original data set. This is accomplished as follows: Scores on X and Y are determined for a sample of individuals. The procedures presented in Chapter 5 are then used to derive a regression equation. This equation can then be applied to individuals outside the original sample to make predictions about their scores on Y from their scores on X. This is accomplished by substituting an individual's X score into the regression equation.

9. Unless the two variables are perfectly correlated, the use of a regression equation to predict scores on Y from scores on X will have some degree of error associated with it. What is needed in order to gain insight into the predictive utility of a regression equation is an index of how much error will occur when predicting Y from X. Such an index is provided by the estimated standard error of estimate. The estimated standard error of estimate estimates in the population the average error that is made across individuals when predicting scores on Y from the regression equation.

Answers to Selected Exercises from Textbook

Exercise Number 44: There are two variables measured for each of the 10 subjects: leader's LPC score and the group's problem-solving performance in minutes. The independent variable is the leader's LPC score, and the dependent variable is the group's problem-solving performance in minutes. Both variables are quantitative and measured on approximately an interval scale. The two variables have been measured on the same individuals and the observations are between-subjects in nature. Therefore, Pearson correlation is appropriate.

The null hypothesis states that the population correlation between these two variables is zero, meaning that there is no linear relationship between the variables. The alternative hypothesis states that the population correlation coefficient is not zero, that there is a linear relationship between the variables.

$$H_0 : \rho = 0$$

$$H_1: \rho \neq 0$$

First, write the formula to see what w need to compute. Since we have the raw scores, we use the computational formula (Equation 5.8).

$$r = \frac{\Sigma XY - \dfrac{(\Sigma X)(\Sigma Y)}{N}}{\sqrt{\left(\Sigma X^2 - \dfrac{(\Sigma X)^2}{N}\right)\left(\Sigma Y^2 - \dfrac{(\Sigma Y)^2}{N}\right)}}$$

Thus we need ΣX, ΣY, ΣX^2, ΣY^2, ΣXY, and N.

Group	X (Leader's LPC score)	Y (Minutes until solution)	X^2	Y^2	XY
1	63	11	3969	121	693
2	68	12	4624	144	816
3	71	15	5041	225	1065
4	65	10	4225	100	650
5	61	6	3721	36	366
6	75	19	5625	361	1425
7	64	9	4096	81	576
8	63	7	3969	49	441
9	70	13	4900	169	910
10	73	7	5329	49	511

$\Sigma X = 673$ $\Sigma Y = 109$ $\Sigma X^2 = 45,499$ $\Sigma Y^2 = 1,335$ $\Sigma XY = 7,453$

Now we can substitute the appropriate values into the formula for the Pearson correlation:

$$r = \cfrac{7{,}453 - \cfrac{(673)(109)}{10}}{\sqrt{\left(45{,}499 - \cfrac{(673)^2}{10}\right)\left(1{,}335 - \cfrac{(109)^2}{10}\right)}}$$

$$= \frac{7{,}453 - 7{,}335.70}{\sqrt{(206.1)(146.9)}} = .67$$

For an alpha level of .05, nondirectional test, and degrees of freedom of N - 2 = 8, the critical values of ρ from Appendix H are ±.632. Our observed Pearson correlation of .67 is greater than the positive critical value of .632, therefore we reject the null hypothesis and conclude that there is a relationship between the leader's LPC and the group's problem-solving performance.

The strength of the relationship is indexed by the square of the correlation coefficient:

$$r^2 = .67^2 = .45$$

The proportion of variability in the group's problem-solving performance that is associated with the leader's LPC score is .45. This represents a strong effect.

The nature of the relationship is indicated by the sign of the correlation coefficient. Here the correlation is positive, indicating that as the leader's LPC score decreases, the group's problem-solving performance also decreases.

Exercise 44 also asks us to compute the regression equation and the estimated standard error of estimate. Here, we are predicting the group's problem-solving performance from the leader's LPC score. First, write out the formula for regression so that we know what we need to compute (Equation 5.9).

$$\hat{Y} = a + bX$$

To compute the regression equation, we need to compute b, the slope, and a, the intercept.

$$b = \frac{SCP}{SS_X}$$

$$a = \overline{Y} - b\overline{X}$$

The sum of the cross-products and the sum of squares for X were already calculated when the correlation coefficient was being computed. Thus,

$$b = \frac{117.3}{206.1} = .5691$$

$$a = 10.90 - (.5691)(67.30) = -27.4$$

The regression equation is:

$$\hat{Y} = -27.4 + .5691\,X$$

To compute the estimated standard error of the estimate, we use Equation 14.6:

$$\hat{s}_{YX} = \sqrt{\frac{SS_Y(1 - r^2)}{N - 2}}$$

$$= \sqrt{\frac{(146.9)\,(1-.67^2)}{10-2}} \; = \; 3.18$$

Exercise 44 also asks us to write the results using the principles discussed in the Method of Presentation section (Note: All standard deviations were computed using Equation 7.2 for a standard deviation estimate in Chapter 7):

Results

A Pearson correlation addressed the relationship between the leader's LPC score ($\underline{M}$ = 67.30, $\underline{SD}$ = 4.79) and the group's problem-solving performance ($\underline{M}$ = 10.90, $\underline{SD}$ = 4.04), using an alpha level of .05. The Pearson correlation was statistically significant, $\underline{r}$(8) = .67, $\underline{p}$ < .05, indicating that these two variables are positively related. The regression equation for predicting the group's problem-solving performance from the leader's LPC score was found to be = -27.4 + .569 X. The estimated standard error of estimate was 3.18.

Exercise 47: In this exercise, there are two variables measured for each of the 10 subjects: GRE-A score and GPA after two years of graduate study. Both variables are quantitative and measured on approximately an interval scale. The two variables have been measured on the same individuals and the observations are between-subjects in nature. Therefore, Pearson correlation is appropriate.

The null hypothesis states that the population correlation between these two variables is zero, meaning that there is no linear relationship between the variables. The alternative hypothesis states that the population correlation coefficient is not zero:

$$H_0 : \rho = 0$$

$$H_1: \rho \neq 0$$

First, write the formula to see what w need to compute. Since we have the raw scores, we use the computational formula (Equation 5.8).

$$r = \frac{\Sigma XY - \dfrac{(\Sigma X)(\Sigma Y)}{N}}{\sqrt{\left(\Sigma X^2 - \dfrac{(\Sigma X)^2}{N}\right)\left(\Sigma Y^2 - \dfrac{(\Sigma Y)^2}{N}\right)}}$$

Thus we need ΣX, ΣY, ΣX^2, ΣY^2, ΣXY, and N.

Student	X (GRE-A score)	Y (GPA)	X^2	Y^2	XY
1	533	3.11	284,089	9.67	1,657.63
2	497	2.89	247,099	8.35	1,436.33
3	612	3.66	374,544	13.40	2,239.92
4	564	3.50	318,096	12.25	1,974.00
5	582	3.29	338,724	10.82	1,914.78
6	476	3.34	226,576	11.16	1,589.84
7	607	3.61	368,449	13.03	2,191.27
8	621	3.74	385,641	13.76	2,322.54
9	590	3.42	348,100	11.70	2,017.80
10	512	2.61	262,144	6.81	1,336.32

$\Sigma X = 5,594$ $\Sigma Y = 33.17$ $\Sigma X^2 = 3,153,372$ $\Sigma Y^2 = 111.178$ $\Sigma XY = 18,680.43$

Now we can substitute the appropriate values into the formula for the Pearson correlation:

$$r = \cfrac{18{,}680.43 - \cfrac{(5{,}594)(33.17)}{10}}{\sqrt{\left(3{,}153{,}372 - \cfrac{(5{,}594)^2}{10}\right)\left(111.178 - \cfrac{(33.17)^2}{10}\right)}}$$

$$= \frac{18{,}680 - 18{,}555.298}{\sqrt{(24{,}088.4)\,(1.158)}} = .75$$

For an alpha level of .05, nondirectional test, and degrees of freedom of $N - 2 = 8$, the critical values of ρ from Appendix H are $\pm.632$. Our observed Pearson correlation of .75 is greater than the positive critical value of .632, therefore we reject the null hypothesis and conclude that there is a relationship between GRE-A scores and GPA after 2 years of graduate study.

The strength of the relationship is indexed by the square of the correlation coefficient:

$$r^2 = .75^2 = .56$$

The proportion of variability in GPA after 2 years of graduate study that is associated with GRE-A scores is .56. This represents a strong effect.

The nature of the relationship is indicated by the sign of the correlation coefficient. Here the correlation is positive, indicating that as subjects' GRE-A scores increase, their GPA in graduates school also increases.

Exercise 44 also asks us to compute the regression equation and the estimated standard error of estimate. Here, we are predicting an individual's GPA after 2 years in grad school from his or her GRE-A scores. First, write out the formula for regression so that we know what we need to compute (Equation 5.9):

$$\hat{Y} = a + bX$$

To compute the regression equation, we need to compute b, the slope, and a, the intercept:

$$b = \frac{SCP}{SS_X}$$

$$a = \overline{Y} - b\overline{X}$$

The sum of the cross-products and the sum of squares for X were already calculated when the correlation coefficient was being computed. Substituting values, we obtain:

$$b = \frac{125.132}{24,088.4} = .0052$$

$$a = 3.32 - (.005)(559.40) = .52$$

Thus, the regression equation is:

$$\hat{Y} = .52 + .005 X$$

Exercise 47 also asks which of the applicants in Set II should be accepted if only students who are likely to maintain a two-year GPA of 3.00 or better are to be admitted. Therefore, we plug each of the seven applicants into the regression equation and predict their two-year GPA. If the predicted GPA is 3.00 or above, we would admit the applicant, but may not admit those below 3.00:

172

$$\hat{Y}_1 = .52 + (.005)(532) = 3.18$$
$$\hat{Y}_2 = .52 + (.005)(478) = 2.90$$
$$\hat{Y}_3 = .52 + (.005)(589) = 3.46$$
$$\hat{Y}_4 = .52 + (.005)(483) = 2.94$$
$$\hat{Y}_5 = .52 + (.005)(527) = 3.16$$
$$\hat{Y}_6 = .52 + (.005)(493) = 2.98$$
$$\hat{Y}_7 = .52 + (.005)(546) = 3.25$$

Based on the regression equation, the predicted two-year grade point average of applicants 1, 3, 5, and 7 are all greater than 3.00, so these applicants should be admitted to the program.

Chapter 15: Chi Square Test

Study Objectives

This chapter covers different applications of what has been traditionally called the "chi square test." After reading the material, you should be able to explain the distinction between the chi square test of independence, the chi square test of homogeneity, and the chi square goodness-of-fit test. You should be able to specify the conditions under which each is applied.

You should know what a contingency table is, including the notion of marginal frequencies. You should understand notation used to refer to contingency tables (e.g., the difference between a 2x2 contingency table and a 4x3 contingency table). You should be able to explain and calculate an observed frequency and an expected frequency. You should be able to calculate a chi square statistic.

You should be able to characterize a sampling distribution of a chi square statistic. You should be able to discuss the theoretical chi square distribution and why it is important for applications of chi square tests.

You should be able to test the null hypothesis in the context of a chi square test. You should be able to calculate and explain Cramer's statistic and the fourfold point correlation coefficient. Why are they computed? You should be able to explore the nature of a relationship by examining discrepancies between observed and expected frequencies.

You should be able to describe situations where the chi square test becomes suspect (e.g., small expected frequencies) and identify analytic alternatives (e.g., Fisher's exact test).

You should be able to discuss issues surrounding the use of quantitative variables in chi square analysis. Under what conditions is it a bad idea to analyze quantitative variables using chi square statistics?

You should be able to describe and apply the chi square goodness of fit.

You should be able to determine the sample size necessary to achieve a given level (e.g., 0.80) of statistical power when designing a study that will use a chi square test. You should be able to explain how alpha, sample size, and the value of the Cramer coefficient in the population influences statistical power for this test.

You should be able to write up the results of an chi square test using APA format as discussed in the Method of Presentation section. You should be able to interpret the results of a chi square test from examples reported in the literature.

Study Tips

A common source of confusion among students is the distinction between the chi square statistic and the chi square distribution. The chi square statistic is an index that is calculated in a set of sample data. It reflects how discrepant the observed and expected frequencies are *in the sample data*. It has the general form:

$$\chi^2 = \Sigma \ \frac{(O_j - E_j)^2}{E_j}$$

Note that if the observed and expected frequencies are identical, the value of χ^2 will be zero and that larger discrepancies will yield a larger χ^2 value, everything else being equal. There are other ways we could calculate a sample statistic to reflect discrepancies between observed and expected frequencies. For example, we could sum the absolute differences between the two:

$$\Sigma \ |O_j - E_j|$$

The reason we prefer the first approach (i.e., the chi square statistic) is that this particular method of calculating the discrepancy between observed and expected frequencies yields a statistic that has a sampling distribution that closely approximates the theoretical chi square distribution. This is not true of the approach using the sum of absolute discrepancies. The theoretical chi square distribution is a distribution that we know a great deal about and we can specify probabilities of obtaining ranges of scores within it. This makes the chi square statistic a very useful index of the discrepancies between observed and expected frequencies because we can use our knowledge of the chi square distribution to make statements about the sampling distribution of the chi square statistic. Keep the distinction between the chi square distribution and the chi square statistic clear. It is a common point of confusion.

Glossary Of Important Terms

Study the terms listed below. Make sure you understand each so that you could explain them to someone else who does not know them.

Chi-Square Test: A statistical procedure used to analyze the relationship between two variables that are qualitative (or quantitative with few values) and which are between subjects in nature.

Contingency Table: A table of numbers in which the columns are the values of one variable and the rows are the values of a second variable. The entry in a given cell of the table is the frequency with which the combination of values defined by the column and row occurs.

Cell: Each unique combination of values of two variables in a contingency table.

Observed Frequencies: Entries within the cells of a contingency table that indicate the number of individuals in the sample who are characterized by the corresponding levels of the variables.

Marginal Frequencies: The sums of the frequencies in the corresponding row or column of a contingency table.

Chi-Square Test of Independence: A chi-square test where the marginal frequencies of both variables under study are random.

Chi-Square Test of Homogeneity: A chi square test where the marginal frequencies are random for one variable and fixed for the other variable.

Expected Frequencies: The frequency of individuals characterized by a given cell in a contingency table under the assumption that the null hypothesis is true.

Chi-Square Statistic: An index of the discrepancy between the observed frequencies and expected frequencies. The index is defined so that it has a sampling distribution that closely approximates a chi square distribution.

Sampling Distribution of the Chi-Square Statistic: A distribution of chi square statistics for all possible random samples of a given size.

Chi-Square Distribution: A theoretical distribution that has known mathematic properties and that permits us to make probability statements about the occurrence of ranges of values. There is a different chi square distribution for each degree of freedom.

Yates' Correction for Continuity: A correction factor that supposedly makes the chi square statistic have a sampling distribution that more closely approximate a chi square distribution. Its use is controversial.

Fisher's Exact Test: A statistical test for evaluating the null hypothesis in a 2X2 contingency table that is more accurate than the traditional chi square test.

Fourfold Point Correlation Coefficient: A common index of the strength of association applied to the relationship between variables in a 2X2 contingency table.

Cramer's Statistic: An index of strength of association when one or both variables in a contingency table have more than two levels.

Goodness-of-Fit Test: A statistical procedure to test the null hypothesis that a population distribution has an *a priori* specified form.

Practice Questions: True-False Format

1. The chi-square test is typically used to analyze the relationship between two variables T when both variables are qualitative in nature (that is, measured on a nominal level).

2. Unlike the previous tests we have considered, the chi-square test analyzes the relationship between variables using frequency information. T

3. The use of frequency information is predicated on the fact that the chi-square test is designed for use with quantitative variables, and it is not appropriate to compute means for variables of this type. F

4. The chi-square test is a parametric statistical test. F

5. Each unique combination of variables in a contingency table is referred to as a cell. T

6. The entries within cells represent the number of individuals in the sample who are characterized by the corresponding levels of the variables and are referred to as marginal frequencies.

7. Cell frequencies are the sum of the frequencies in the corresponding row or column.

8. If people in a study on gender and political party identification were selected for participation without regard to gender or political party identification, the marginal frequencies for both of these variables would be free to vary, or random.

9. If people in a gender and political party identification study were selected so that it consisted of a specified number of individuals of each gender (for instance, 100 males and 100 females) who were then classified according to political party identification, the marginal frequencies for the party identification variable would still be random, but the marginal frequencies for the gender variable would be fixed.

10. When the marginal frequencies of both variables under study are random, the chi-square test is known as the chi-square test of dependence.

11. When the marginal frequencies are random for one variable and fixed for the other, the analytical procedures are identical but the test is referred to as the chi-square test of heterogeneity.

12. The logic underlying the chi-square test focuses on the concept of expected frequencies and how they compare to observed frequencies.

13. The null hypothesis for a chi-square test states that the variables of interest are unrelated in the population, while the alternative hypothesis states that there is a relationship between the two variables in the population.

14. Because the relationship between two variables can take a number of different forms, the alternative hypothesis for the chi-square test, as with the F test in analysis of variance, is directional in nature.

15. Application of the chi-square test requires computation of an expected frequency for each cell under the assumption of no relationship between the two variables.

16. In order to calculate an expected frequency for a given cell, you divide the total of the column (the column marginal frequency) in which it appears by the total number of observations and then multiply this by the total of the row (the row marginal frequency) in which the cell appears.

17. The chi-square statistic is an index that reflects the overall difference between the observed and the expected frequencies.

18. If two variables are unrelated in the population, the population value of chi-square will equal 1.0.

19. Because of sampling error, a chi-square computed from sample data might be less than 0 even when the null hypothesis is true.

20. There are different chi-square distributions depending on the degrees of freedom associated with them.

21. Since all discrepancies from the expected frequencies are reflected in the upper tail of the chi-square distribution (as defined by the critical value), the chi-square test is nondirectional.

22. The chi-square test is based on the assumption that the observations are independently sampled from the population of all possible observations.

23. The chi-square test is based on the assumption that the expected frequency for each cell is nonzero.

24. Although the issue is controversial, statisticians generally recommend the use of the correction for continuity.

25. When both of the variables under study have only two levels, the sampling distribution of the chi-square statistic corresponds more closely to a chi-square distribution than when one or both variables have more than two levels.

26. Yates' correction for continuity involves subtracting .5 from the absolute value of $O_j - E_j$ before these quantities are squared, divided by E_j, and summed across cells.

27. Research has found that Fisher's Exact Test, an alternative method for testing the relationship between two variables in 2 X 2 tables, is less powerful than the chi-square test.

28. Probably the most common index of the strength of the relationship between two variables in a contingency table is a measure known as the fourfold point correlation coefficient (as it is called when applied to the relationship between variables with two levels each) or Cramér's statistic (as it is called when one or both variables have more than two levels).

29. The fourfold point correlation coefficient and Cramér's statistic can range from -1.00 to 1.00.

30. Conceptually, a large value of V indicates a tendency for particular categories of one variable to be associated with particular categories of the other variable.

31. The test of the chi-square statistic applies to the sample as a whole and provides useful information as to which cells are responsible for rejecting the null hypothesis.

32. Because the chi-square test is typically used to analyze the relationship between two qualitative variables, it cannot be used when one or both variables are quantitative.

33. The question addressed by the goodness-of-fit test is whether a distribution of frequencies across categories for a variable in a population are distributed in a specified manner.

34. For the chi-square goodness-of-fit test, statisticians generally recommend that the lowest expected frequency of any category be somewhere between 5 and 10 if only two categories are involved and about 5 if more than two categories are involved.

Answers to True-False Items

1. T	11. F	21. T	31. F
2. T	12. T	22. T	32. F
3. F	13. T	23. F	33. T
4. F	14. F	24. F	34. T
5. T	15. T	25. F	
6. F	16. T	26. T	
7. F	17. T	27. F	
8. T	18. F	28. T	
9. T	19. F	29. F	
10. F	20. T	30. T	

Practice Questions: Short Answer

1. When is Pearson correlation typically used to analyze the relationship between two variables?

2. How is the chi-square test different from previously considered statistical tests?

3. Distinguish between the chi-square test of independence and the chi-square test of homogeneity.

4. Discuss the concept of expected frequencies.

5. Summarize the steps involved in computing expected frequencies.

6. What is the chi-square statistic?

7. Discuss the sampling distribution of the chi-square statistic and the chi-square distribution.

8. What are the assumptions of the chi-square test?

9. What is Fisher's Exact Test?

10. How is the strength of the relationship between two variables in a contingency table evaluated?

11. Why is it necessary to conduct follow-up tests in a chi square analysis in order to determine the nature of the relationship?

12. What are the advantages and disadvantages of applying the chi-square test to quantitative variables?

13. What is the goodness-of-fit test?

Answers to Short Answer Questions

1. The chi-square test is typically used to analyze the relationship between two variables when:(1) both variables are qualitative in nature (that is, measured on a nominal level); (2) the two variables have been measured on the same individuals, and (3) the observations on each variable are between-subjects in nature.

2. Unlike the previous statistical tests that we have considered, the chi-square test analyzes relationships between variables using frequency information. The use of frequency information is predicated on the fact that the chi-square test is designed for use with qualitative variables, and it is not appropriate to compute means for variables of this type.

3. When the marginal frequencies of both variables under study are random, or free to vary, the test is referred to as the chi-square test of independence. When the marginal frequencies are random for one variable and fixed for the other, the analytical procedures are identical but the test is referred to as the chi-square test of homogeneity.

4. The logic underlying the chi-square test focuses on the concept of expected frequencies. An expected frequency is the number of people you would expect to observe in a given cell, assuming the null hypothesis of no relationship is true. If expected frequencies deviate substantially from observed frequencies, then the null hypothesis is called into question.

5. The steps involved in the calculation of expected frequencies can be summarized as follows, based on the frequencies observed in a contingency table: (1) For the cell in question, divide the total of the column (the column marginal frequency) in which it appears by the total number of observations (2) multiply this value by the total of the row (the row marginal frequency) in which the cell appears.

6. The chi-square statistic is an index that reflects the overall difference between the observed and the expected frequencies. It is defined as follows:

$$\chi^2 = \Sigma \ \frac{(O_j - E_j)^2}{E_j}$$

where χ^2 is the value of the chi-square statistic, O_j is the observed frequency for cell j, E_j is the expected frequency for cell j, and the summation is across all cells.

7. A sampling distribution of the chi-square statistic can be formed by computing sample chi-square statistics for all possible random samples of a given size. This distribution closely approximates, under certain conditions, a theoretical distribution called the chi-square distribution. There are different chi-square distributions depending on the degrees of freedom associated with them. For the chi-square test that we are considering, $df = (r - 1)(c - 1)$, where r is the number of levels of the row variable and c is the number of levels of the column variable. Analogous to the previous sampling distributions we have considered, probability statements can be made with respect to scores in the chi-square distribution. It is therefore possible to set an alpha level and define a critical value such that if the null hypothesis is true, the probability of obtaining a chi-square value larger than that critical value is less than alpha. Since all discrepancies from the expected frequencies are reflected in the upper tail of the chi-square distribution (as defined by the critical value), the chi-square test is, by its nature, nondirectional.

8. The chi-square test is based on several assumptions. These ensure that the sampling distribution of the chi-square statistic approximates a chi-square distribution. Specifically, the following is assumed (1) the observations are independently and randomly sampled from the population of all possible observations; (2) the expected frequency for each cell is nonzero.

9. Methods other than the chi-square statistic have been suggested for testing the relationship between variables in 2 X 2 tables. One of the more popular alternatives is Fisher's Exact Test. Research has generally found the Fisher exact method to be preferable to the chi-square test (although they converge to the same result as N increases).

10. Probably the most common index of the strength of association is a measure known as the fourfold point correlation coefficient (as it is called when applied to the relationship between variables with two levels each) or Cramér's statistic (as it is called when one or both variables have more than two levels). The fourfold point correlation/Cramér's statistic can range from 0 to 1.00, where a value of 0 indicates no relationship and a value of 1.00 indicates a perfect relationship.

11. If the null hypothesis of no relationship is rejected, then additional steps are required to discern more fully the nature of the relationship. The test of the chi-square statistic applies to the data taken as a whole and provides no useful information as to which cells are responsible for rejecting the null hypothesis. Just as the Tukey HSD test can be applied to break down the overall relationship following a statistically significant analysis of variance, comparable tests can be applied following a statistically significant chi-square test.

12. While the chi-square test is typically used to analyze the relationship between two qualitative variables, it can also be applied when one or both variables are quantitative. A common procedure involves classifying scores on a quantitative variable into a small number of groups before applying the chi-square test. When possible, however, it is usually preferable to analyze quantitative variables with the parametric tests discussed in prior chapters than with the chi-square test. Parametric tests are usually preferred over the chi-square test because they tend to be more powerful. The power of the chi-square test is further reduced when quantitative variables are collapsed into categories because considerable information is likely to be lost by placing individuals with different scores into the same group. While this approach implicitly assumes that all individuals assigned to a given category are equivalent on the underlying dimension, this may not in fact be the case. An advantage of the chi-square approach, however, is that a quantitative variable need be measured on only an ordinal level as opposed to the approximately interval level required for parametric tests.

13. The question addressed by the goodness-of-fit test is whether a distribution of frequencies across categories for a variable in a population are distributed in a specified manner. A common use of the goodness-of-fit test is to test whether frequencies in the population are evenly distributed across the categories under study.

Answers to Selected Exercises from Textbook

Exercise Number 28: The null hypothesis is that the population represented by the sample does not differ from the general population in the distribution of ratings of evening news programs. The alternative hypothesis is that the population represented by the sample does differ from the general population in the distribution of ratings of evening news programs.

We compute the expected frequencies by multiplying the relative frequencies of each category by the overall sample size. First, write the formula (Equation 15.6).

$$E_j = (rf_j)(N)$$

The first table in Exercise 28 provides the relative frequencies, and N is the 1,000 students sampled. Therefore, the expected frequencies are as follows.

$$\text{ABC:} \quad E_j = (.20)(1,000) = 200.00$$

$$\text{NBC:} \quad E_j = (.23)(1,000) = 230.00$$

$$\text{CBS:} \quad E_j = (.21)(1,000) = 210.00$$

$$\text{NP:} \quad E_j = (.36)(1,000) = 360.00$$

The observed frequencies are in the second table in Exercise 28. We can now place the observed and expected values into the computational table, and finish the columns of computation:

	O	E	O-E	$(O-E)^2$	$(O-E)^2/E$
ABC	117	200	-23.00	529	2.64
NBC	252	230	22.00	484	2.10
CBS	240	210	30.00	900	4.29
No Preference	331	360	-29.00	841	2.34

$$\chi^2 = 11.37$$

The critical value of chi-square for an alpha level of .05 and 3 degrees of freedom is 7.815. Since the observed chi-square value of 11.37 is greater than 7.815, we reject the null hypothesis and conclude that the evening news program preference for the population of college students represented by this sample differs from that reflected in the national ratings. Examination of the O - E and the $(O - E)^2$ values indicate that college students are more likely than expected to prefer the CBS and NBC evening news programs, and less likely than expected to prefer the ABC evening news program or to have no evening news program preference.

Exercise 46: For Exercise 46, we have two variables, both qualitative in nature. The variables have been measured on the same individuals and the observations on each variable are between-subjects. Thus, chi-square is appropriate for the analysis of the relationship between the two variables.

The null hypothesis states that the variables are unrelated in the population. The alternative hypothesis states that there is a relationship between the two variables in the population. The alternative hypothesis can take a number of different forms and is thus nondirectional.

H_0: Gender and role of the central character are unrelated in the population.

H_1: Gender and role of the central character are related in the population.

We begin by computing the expected frequencies. First, write the formula for the expected frequencies (Equation 15.1):

$$E_j = \frac{CMF_j}{N} (RMF_j)$$

Using this formula, we compute an expected frequency for each cell:

Male authority: $E_j = (158/315)(252) = 126.00$

Male user: $E_j = (157/315)(252) = 126.00$

Female authority: $E_j = (158/315)(63) = 31.50$

Female user: $E_j = (158/315)(63) = 31.50$

Now, place the observed frequencies from the contingency table in Exercise 46 and the expected frequencies from our computations into a table and finish the chi-square computations:

	O	E	O-E	$(O-E)^2$	$(O-E)^2/E$
Male authority	138	126.0	12.0	144.00	1.14
Male user	114	126.0	-12.0	144.00	1.14
Female authority	20	31.5	-11.5	132.25	4.20
Female user	43	31.5	11.5	132.25	4.20

$$\chi^2 = 10.68$$

The critical value of chi-square from Appendix J for an alpha level of .05 and $(r - 1)(c - 1) = 1$ degree of freedom is 3.814. Since the observed chi-square value of 10.68 is greater than 3.814, we reject the null hypothesis and conclude that their is a relationship between gender of the central character and the role portrayed by that character.

The strength of the relationship is indicated by the fourfold point correlation coefficient. First, write the formula (Equation 15.5).

$$V = \sqrt{\dfrac{\chi^2}{N(L-1)}}$$

The observed chi-square is 10.68, N is 315, and L is 2:

$$V = \sqrt{\dfrac{10.68}{315\,(2-1)}} = .18$$

The fourfold point correlation coefficient is moderate. There is only a moderate tendency for the a given category of one variable to be associated with particular categories of the other variable.

The nature of the relationship can be seen by examining the O - E column of the chi-square computational table. Examination of the O - E column indicates female authority roles are less likely than expected and female user roles are more likely than expected. Also in the O - E column, we see that male authority roles are more likely than expected and male user roles are less likely than expected. Thus, female central characters are more likely than male central characters to be portrayed as users as opposed to authorities.

Exercise 46 also asks us to write the results using the principles discussed in the Method of Presentation section.

Results

The relationship between gender of the central character and the role portrayed by that character was analyzed using a chi-square test, with an alpha level of .05. The chi-square was statistically significant, χ^2 (1, $\underline{N}$ = 315) = 10.68, $\underline{p}$ < .01. The strength of the relationship, as indexed by the fourfold point correlation coefficient, was .18. This reflects the fact that female central characters are more likely than male central characters to be portrayed as users as opposed to authorities.

Chapter 16: Nonparametric Statistical Tests

Study Objectives

This chapter covers a wide range of nonparametric statistical methods. After reading the material, you should be able to characterize nonparametric tests in relation to parametric tests. You should be able to explain the conditions where we might prefer nonparametric tests. You should be able to determine how to rank order a set of scores, and what to do in the case of ties. You should be able to describe the rank transformation approach and how it draws upon traditional parametric methods of analysis. You should be able to describe what is meant by an outlier resistant statistical test.

For each of the non-parametric methods presented, you should be able to identify its parametric counterpart and know how to address the three fundamental questions: (1) is there a relationship between the independent and dependent variables, (2) what is the strength of the relationship, and (3) what is the nature of the relationship.

You should be able to write up the results of the various nonparametric tests using APA format as discussed in the Method of Presentation section. You should be able to interpret the results of these analyses from examples reported in the literature.

Study Tips

This chapter covers a great deal of material in a short amount of space. Most of the material is straightforward, if you have mastered previous chapters in the textbook. However, there is a large amount of information. What is most important from a "big picture" perspective is that you are able to accurately identify the parametric and nonparametric counterparts to each other (e.g., the Wilcoxon rank sum test is the counterpart to the independent groups t test) and that you know how to answer our three basic questions about relationships for any given nonparametric test.

Glossary Of Important Terms

Study the terms listed below. Make sure you understand each so that you could explain them to someone else who does not know them.

Nonparametric Tests: A class of statistical methods for testing the viability of a null hypothesis while making no assumptions about the distribution of scores in the population (hence, they are often called *distribution free* tests).

Correction Term: In the context of nonparametric tests, it is an adjustment to a formula for a test statistic in order to adjust for the presence of ties.

Outlier Resistant: When the conclusions of a statistical test are not influenced by the presence of outliers.

Rank Transformation Approach: A class of statistics for testing the viability of a null hypothesis based on converting a set of scores on a variable to ranks, and then analyzing these rank scores using traditional parametric formulas.

Wilcoxon Rank Sum Test: The nonparametric counterpart of the independent groups t test. It is typically used to analyze the relationship between two variables when (1) scores on the dependent variable are in the form of ranks, (2) the independent variable is between-subjects in nature (it can be either qualitative or quantitative), and (3) the independent variable has two and only two levels.

Mann-Whitney U Test: An alternative to the Wilcoxon rank sum test and used under the same conditions as the Wilcoxon Rank Sum test but when the sample size is small.

R Statistic: A test statistic used in the Wilcoxon rank sum test.

U Statistic: A test statistic used in the Mann-Whitney U test.

Glass Rank Biserial Correlation Coefficient: A coefficient that measures the strength of the relationship between two variables in the Wilcoxon rank sum test or Mann-Whitney U test.

Wilcoxon Signed-Rank Test: The nonparametric counterpart of the correlated groups t test. It is typically used to analyze the relationship between two variables when (1) scores on the dependent variable are in the form of ranks, (2) the independent variable is within-subjects in nature (it can be either qualitative or quantitative), and (3) the independent variable has two and only two levels.

T Statistic: A test statistic used in the Wilcoxon signed rank test.

Matched-Pairs Rank Biserial Correlation Coefficient: A statistical index of the strength of the relationship for the Wilcoxon signed-rank test.

Kruskal-Wallis Test: The nonparametric counterpart of one-way between-subjects analysis of variance. It is typically used to analyze the relationship between two variables when (1) scores on the dependent variable are in the form of ranks, (2) the independent variable is between-subjects in nature (it can be either qualitative or quantitative), and (3) the independent variable has three or more levels.

Epsilon-Squared: A statistical index used to measure the strength of the relationship for the Kruskal-Wallis test.

Dunn Procedure: A statistical procedure for determining the nature of the relationship in the context of the Kruskal-Wallis test.

Friedman Analysis of Variance by Ranks: The nonparametric counterpart of one-way repeated measures analysis of variance. It is typically used to analyze the relationship between two variables when (1) scores on the dependent variable are in the form of ranks across conditions for each subject, (2) the independent variable is within-subjects in nature (it can be either qualitative or quantitative), and (3) the independent variable has three or more levels.

Spearman Rank-Order Correlation: The nonparametric counterpart of Pearson correlation. It is typically used to analyze the relationship between two variables when (1) scores on both variables are in the form of ranks, (2) the two variables have been measured on the same individuals, and (3) the observations on each variable are between-subjects in nature.

Practice Questions: True-False Format

1. Statistical tests that require assumptions about the distribution of scores in the populations from which the samples are selected are called parametric statistical tests.

2. Nonparametric statistical tests are a class of statistics that, in general, make fewer distributional assumptions than parametric statistics.

3. Nonparametric tests can not be used to analyze quantitative variables that are measured on an ordinal level.

4. Rather than comparing groups in terms of means (as is the case with t tests and analysis of variance), the nonparametric procedures compare groups in terms of medians or other features of a distribution.

5. One factor that is important in deciding whether to use a parametric or non-parametric test is the decision about what aspect of the distribution of the independent variable one wishes to focus on.

6. If one wishes to compare medians across groups, then a parametric test should be pursued.

7. There are two major types of nonparametric tests -- rank tests and sign tests.

8. In cases where a tie occurs in rank scores, the tied scores are discarded.

9. Rank-order tests assume that qualitative variables are continuous in nature.

10. The most common approach to dealing with tied scores is to rank the scores, assign the tied scores the average of the ranks involved, and then to introduce a correction term into the formula for the test statistic to adjust for the presence of ties.

11. As long as the number of ties is minimal, the test statistics will not be affected greatly by tied ranks.

12. A weakness of many nonparametric statistical tests is their relative lack of sensitivity to outliers.

13. The rank transformation approach involves converting a set of scores on a variable to ranks, and then analyzing these rank scores using the traditional parametric formulas.

14. The Wilcoxon rank sum test and the Mann-Whitney U test are the parametric counterparts of the independent groups t test.

15. The sum of the ranks for a given group, R_j, is known as the R statistic.

16. In the Wilcoxon rank sum test, when the sample sizes of both groups are 10 or greater, the shape of the sampling distribution of R_j approximates a normal distribution, which allows us to convert the R statistic to a z score.

17. The Wilcoxon rank sum test is not applicable when the sample sizes of both groups are 10 or greater.

192

18. When one or both sample sizes are smaller than 10 and the independent variable has two and only two values and is between-subjects in nature, then the data can be analyzed using the Mann-Whitney U test.

19. Regardless of whether the Wilcoxon rank sum test or the Mann-Whitney U test is used, the strength of the relationship between the two variables cannot be determined.

20. The Wilcoxon signed-rank test is the nonparametric counterpart of one-way repeated measures analysis of variance.

21. For the Wilcoxon signed-rank test, in order for the null hypothesis to be rejected, the observed value of T must be equal to or less than the critical value.

22. The Kruskal-Wallis test is the nonparametric counterpart of one-way between subjects analysis of variance.

23. In a Kruskal-Wallis test, if the null hypothesis of no relationship between the independent variable and the dependent variable is true, we would expect the mean ranks for the three conditions to be different, within the constraints of sampling error.

24. The strength of the relationship for the Kruskal-Wallace test can be measured with eta-squared.

25. Analogous to analysis of variance, rejection of the null hypothesis when a nonparametric test is applied to more than two groups tells us only that at least two of the groups differ, but does not indicate the nature of the relationship.

26. Friedman analysis of variance by ranks is the nonparametric counterpart of one-way repeated measures analysis of variance.

27. Friedman analysis of variance by ranks involves first rank ordering the means for each experimental group.

28. Spearman rank-order correlation, more simply known as Spearman correlation, is a nonparametric counterpart of Pearson correlation.

29. The Spearman rank-order correlation coefficient ranges from 0 to +1.00 and is interpreted in the same manner as the coefficient of determination.

30. The strength of the relationship for Spearman correlation is indicated by the sign of the

correlation coefficient and the nature of the relationship is indicated by the magnitude of the correlation coefficient.

Answers to True-False Items

1. T	11. T	21. T
2. T	12. F	22. T
3. F	13. T	23. F
4. T	14. F	24. F
5. F	15. T	25. T
6. F	16. T	26. T
7. T	17. F	27. F
8. F	18. T	28. T
9. F	19. F	29. F
10. T	20. F	30. F

Practice Questions: Short Answer

1. Distinguish between parametric and nonparametric statistics.

2. How are tied values dealt with when applying a nonparametric test?

3. What is meant by the phrase "outlier resistant"?

4. What is the rank transformation approach?

5. Under what conditions are the Wilcoxon rank sum test and the Mann-Whitney U test used to analyze the relationship between two variables?

6. Under what conditions is the Wilcoxon signed-rank test used to analyze the relationship between two variables?

7. Under what conditions is the Kruskal-Wallis test used to analyze the relationship between two variables?

8. When the null hypothesis is rejected using a nonparametric statistic applied to more than two groups, how is the nature of the relationship between the independent and dependent variables assessed?

9. Under what conditions is the Friedman analysis of variance by ranks used to analyze the relationship between two variables?

10. Under what conditions is the Spearman rank-order correlation used to analyze the relationship between two variables?

Answers to Short Answer Questions

1. Parametric statistical tests require assumptions about the distributions of scores in the populations from which samples are selected (e.g., normality of distributions and homogeneity of variances). Nonparametric statistical tests are a class of statistics that, in general, make fewer distributional assumptions than parametric statistics. Because of this, they are often called distribution-free tests. Nonparametric tests can be used to analyze quantitative variables that are measured on an ordinal level. In contrast, parametric statistical tests are most appropriately applied to quantitative variables that are measured on a level that at least approximates interval level characteristics.

2. In cases where a tie occurs, the tied scores are assigned the average of the ranks involved.

3. A benefit of many nonparametric statistical tests is their relative lack of sensitivity to outliers. Because data are converted to ranks, outliers do not play the kind of havoc that they can on such statistics as means, correlations, and standard deviations. Because of this property, nonparametric tests are often said to be outlier resistant.

4. The rank transformation approach involves converting a set of scores on a variable to ranks, and then analyzing these rank scores using the traditional parametric formulas.

5. The Wilcoxon rank sum test and the Mann-Whitney U test are the nonparametric counterparts of the independent groups t test. They are typically used to analyze the relationship between two variables when (1) scores on the dependent variable are in the form of ranks, (2) the independent variable is between-subjects in nature (it can be either quantitative or qualitative), and (3) the independent variable has two and only two levels. When one or both sample sizes are smaller than 10, application of the Wilcoxon rank sum test is not appropriate and the Mann-Whitney U test must be used instead.

6. The Wilcoxon signed-rank test is the nonparametric counterpart of the correlated groups t test. It is used to analyze the relationship between two variables when (1) scores on the dependent variable are in the form of ranked differences, (2) the independent variable is within-subjects in nature, and (3) the independent variable has two and only two levels.

7. The Kruskal-Wallis test is the nonparametric counterpart of one-way between-subjects analysis of variance. It is typically used to analyze the relationship between two variables when (1) scores on the dependent variable are in the form of ranks, (2) the independent variable is between-subjects in nature (it can be either quantitative or qualitative), and (3) the independent variable has three or more levels.

8. Analogous to analysis of variance, rejection of the null hypothesis when a nonparametric test is applied to more than two groups tells us only that at least two of the k groups differ in ranks, but does not indicate the nature of these differences. Several procedures have been proposed for analyzing the nature of the relationship following a statistically significant Kruskal-Wallis test. For example, the Dunn procedure involves using the Wilcoxon rank sum test or the Mann-Whitney U test to compare the mean ranks for all possible pairs of conditions, just as the Tukey HSD test was applied to all possible pairs of means following a statistically significant analysis of variance.

9. The Friedman analysis of variance by ranks is the nonparametric counterpart of one-way repeated measures ANOVA. It is used to analyze the relationship between two variables when (1) scores on the dependent variable are in the form of ranks across conditions for each subject, (2) the independent variable is within-subjects in nature (it can be either quantitative or qualitative), and (3) the independent variable has three or more levels.

10. Spearman rank-order correlation. more simply known as Spearman correlation, is a nonparametric counterpart of Pearson correlation. It is typically used to analyze the relationship between two variables when (1) scores on both variables are in the form of ranks, (2) the two variables have been measured on the same individuals, and (3) the observations on each variable are between-subjects in nature.

Answers to Selected Exercises from Textbook

Exercise Number 7: Exercise 7 asks us to compute a nondirectional Wilcoxon rank sum text and analyze the nature of the relationship. First , write the formula for the Wilcoxon rank sum text so we know what we need to compute (Equation 16.3):

$$z = \frac{|R_j - E_j| - 1}{\sigma_R}$$

We need R_j, E_j, and σ_R. We first do the computations for the sum of the ranks for each group. We can set up our computations in a table:

Individual	Own Car?	GPA	Rank
1	Y	3.10	21
2	Y	2.50	9
3	Y	2.75	14
4	Y	2.96	18
5	Y	3.50	25
6	Y	2.10	3
7	Y	2.30	7
8	Y	2.15	4
9	Y	3.30	23
10	Y	3.70	27
11	Y	2.90	17
12	Y	3.90	29
13	Y	2.70	13
14	Y	2.64	11
15	Y	1.90	1
16	N	2.20	6
17	N	2.40	8
18	N	2.60	10
19	N	2.80	16
20	N	3.20	22
21	N	3.00	20
22	N	2.90	5
23	N	2.98	19
24	N	2.77	15
25	N	2.66	12
26	N	2.00	2
27	N	3.40	24
28	N	3.60	26
29	N	4.00	30
30	N	3.80	28

$$R_1 = 21 + 9 + \ldots + 11 + 1 \quad = \quad 222$$
$$R_2 = 6 + 8 + \ldots + 30 + 28 \quad = \quad 243$$

We rank the scores across all individuals, then sum the ranks for each group. The sum of the ranks for group 1 (owns car) is 222 and the sum of the ranks for group 2 (does not own car) is 243.

Now we need the expected rank sum for each group (Equation 16.1).

$$E_j = \frac{n_j(n_1 + n_2 + 1)}{2}$$

$$= \frac{15(15 + 15 + 1)}{2} = 232.50$$

The standard deviation is computed using Equation 16.2:

$$\sigma_R = \sqrt{\frac{n_1 n_2(n_1 + n_2 + 1)}{12}}$$

$$= \sqrt{\frac{(15)(15)(15 + 15 + 1)}{12}} = 17.05$$

Now we can substitute values into the z formula.

$$z = \frac{|\,R_j - E_j\,| - 1}{\sigma_R}$$

$$= \frac{|243 - 232.50| - 1}{24.109} = .39$$

For an alpha of .05, nondirectional text, the critical values of z from Appendix B are ±1.96. Since the observed z value of .39 is neither less than -1.96 nor greater than +1.96, we fail to reject the null hypothesis of no relationship between car ownership and performance in school.

Exercise 9: Exercise 9 asks us to do a Mann-Whitney U test and specify the nature of the relationship. First, write the formula for the Mann-Whitney U test so we know what we need to compute (Equations 16.4, 16.5).

$$U_1 = n_1 n_2 + \frac{n_1(n_1 + 1)}{2} - R_1$$

$$U_2 = n_1 n_2 + \frac{n_2(n_2 + 1)}{2} - R_2$$

We need to compute the sum of the ranks for group 1 and for group 2. We can organize the computations in a table:

Individual	Gender	Attitude	Rank
1	M	50	2.5
2	M	48	1.0
3	M	74	12.0
4	M	56	5.0
5	M	59	6.0
6	M	65	9.0
7	M	67	11.0
8	M	82	14.0

199

Individual	Gender	Attitude	Rank
9	F	52	4.0
10	F	50	2.5
11	F	76	13.0
12	F	63	8.0
13	F	61	7.0
14	F	66	10.0
15	F	86	16.0
16	F	84	15.0

$$R_1 = 2.5 + 1 + ... + 11 + 14 = 60.5$$
$$R_2 = 4 + 2.5 + ... + 16 + 15 = 75.5$$

The sum of ranks for males is 60.5, and that for females is 75.5. We can substitute these values into the Mann-Whitney U formula:

$$U_1 = (8)(8) + \frac{8(8+1)}{2} - 60.5 = 39.50$$

$$U_2 = (8)(8) + \frac{8(8+1)}{2} - 75.5 = 24.50$$

Since 24.50 is less than 39.50, U equals 24.50. The critical value of U in Appendix K for an alpha level of .05, nondirectional test, with $n_1 = n_2 = 8$ is 13. Since the observed U value of 24.50 is not equal to or less than 13, we fail to reject the null hypothesis of no relationship between gender and attitude toward a newly developed male birth control pill.

Exercise 15: Exercise 15 asks us to conduct a Friedman analysis of variance by ranks. First, write the formula to see what we need to compute (Equation 16.12).

$$\chi_r^2 = \frac{12 \Sigma R_j^2}{N k (k+1)} - 3 N (k+1)$$

We need the sum of the ranks for each condition.. The data are in Exercise 15. We need to rank each individual across the three brands.

Repairman	Rank A	Rank B	Rank C
1	1	2	3
2	1	2	3
3	1	3	2
4	1	2	3
5	3	2	1
6	3	1	2
7	1	2	3
8	1	2	3
9	1	2	3
10	1	2	3
	$R_1 = 14$	$R_2 = 20$	$R_3 = 26$

We are now ready to substitute these values into the Friedman ANOVA by ranks formula:

$$\chi_r^2 = \left(\frac{12 \; \Sigma \; 14^2 + 20^2 + 26^2}{(10)(3)(3+1)} \right) - (3)(10)(3+1)$$

$$= 127.20 - 120 = 7.20$$

For an alpha level of .05, and k - 1 = 3 - 1 = 2 degrees of freedom, the critical value of for the chi-square table in Appendix J is 5.991. Since the observed Friedman ANOVA by ranks value of 7.20 is greater than 5.991, we reject the null hypothesis and conclude that there is a relationship between the ranks of brand of television and ratings of the quality of tubes.

Exercise 17: Exercise 17 asks us to conduct a nondirectional Spearman correlation. First, write the formula so we know what we need to compute (Equation 16.14).

$$r_s = 1 - \frac{6\Sigma D^2}{N(N^2 - 1)}$$

We need the ΣD^2, the sum of the squared differences between the ranks:

City	Ranks of crime rate	Ranks of size of police force	Difference (D)	D^2
1	3	4	-1	1
2	15	16	-1	1
3	18	5	13	169
4	1	2	-1	1
5	8	12	-4	16
6	14	15	-1	1
7	13	6	7	49
8	2	1	1	1
9	7	11	-4	16
10	12	7	5	25
11	16	18	-2	4
12	6	10	-4	16
13	11	9	2	4
14	17	14	3	9
15	10	8	2	4
16	4	17	-13	169
17	9	13	-4	16
18	5	3	2	4
				$\Sigma D^2 = 506$

Now we can substitute these values into the formula for the Spearman correlation:

$$r_s = 1 - \frac{6\ (506)}{18\ (18^2 - 1)}$$

$$= 1 - \frac{3,036}{5,814} = .48$$

Since N is 18, we use Appendix M to test the Spearman correlation coefficient. The critical values for the Spearman correlation for an alpha level of .05, nondirectional test, and N = 18 degrees of freedom are ±.475. Since the observed Spearman correlation coefficient of +.48 is greater than +.475, we reject the null hypothesis and conclude that there is a relationship between the ranks of crime rate in cities and the ranks of the size of the city's police force.

The strength of the relationship as indicated by the magnitude of the correlation coefficient is .48. The nature of the relationship is indicated by the sign of the correlation coefficient. In this case, the positive correlation coefficient indicates that as the rank of crime rate in a city increases, so too does the rank of the size of the city's police force.

Exercise 42: Exercise 42 asks us to conduct a nondirectional Wilcoxon signed-rank test and draw a conclusion. The behavioral measure under study is a quantitative variable measured on an ordinal level. The independent variable is the group encounter session and the dependent variable is the subject's self-esteem. The independent variable is within-subjects.

Begin by computing the difference between scores in the two conditions for each individual. The differences are then ranked from smallest to largest, ignoring the sign of the difference, and ignoring differences of zero.

Individual	Before Session	After Session	Difference	Rank
1	60	67	7	6
2	38	42	-4	4
3	43	37	6	5
4	39	42	-3	3
5	34	26	8	7
6	30	39	-9	8
7	54	43	11	9
8	48	50	-2	2
9	72	71	1	1
10	49	62	-13	10

Now separately sum the ranks for the positive differences (28) and the ranks for the negative differences (27). The expected value is $(28 + 27)/2 = 27.5$. The value of T is the smaller of the sums of the ranks for the differences, or 27 in our case.

The critical value of the Wilcoxon signed-rank test presented in Appendix L for an N of 10, with an alpha level of .05, nondirectional test, is 8. The observed value of the Wilcoxon signed-rank test is 27, which is not equal to or less than the critical value of 8. Therefore, we fail to reject the null hypothesis of no relationship between time of assessment and the subject's self-esteem.

Exercise 44 also asks us to write the results using the principles given in the Method of Presentation section:

Results

A Wilcoxon signed-rank test compared individuals' self-esteem before versus after participating in the group encounter session, using an alpha level of .05. The rank sums were nonsignificantly different, $N = 10$, $T = 16.5$, ns.

Exercise 43: Exercise 43 asks us to compute the Kruskal-Wallis test and draw a conclusion. We have a quantitative dependent variable measured on a ratio level. The independent variable is the affective content of the word, with three levels: positive, negative, and neutral. The dependent variable is the number of words correctly recalled. Different subjects were used for each of the three levels of the independent variable, so we have a between-subjects design.

Begin by writing the formula for the Kruskal-Wallis test to see what we need to compute (Equation 16.10).

$$H = \frac{12}{N(N+1)} \Sigma \frac{R_j^2}{n_j} - 3(N+1)$$

We need the sum of the ranks for each group. We can set the data up in a table:

Individual	Word Content	Learning Score	Rank
1	Pos	20	14
2	Pos	8	3
3	Pos	2	1
4	Pos	17	12
5	Pos	19	13
6	Neg	15	10
7	Neg	10	5
8	Neg	9	4
9	Neg	7	2
10	Neg	11	6
11	Neut	16	11
12	Neut	21	15
13	Neut	14	9
14	Neut	13	8
15	Neut	12	7

The sum of the ranks for the positive word group is 43, for the negative word group is 27, and for the neutral word group is 50. Now we can substitute the appropriate values into the formula for the Kruskal-Wallis test:

$$H = \left(\frac{12}{15\,(15+1)} \right) \left(\Sigma \frac{43^2}{5} + \frac{27^2}{5} + \frac{50^2}{5} \right) - 3\,(15+1)$$

$$= \;(.05)(1{,}015.6) - 48 \;=\; 2.78$$

For an alpha level of .05 and k - 1 = 2 degrees of freedom the critical value of H from the chi-square table in Appendix J is 5.991. Since the observed Kruskal-Wallis value of 2.78 is less than 5.991, we fail to reject the null hypothesis.

Exercise 43 also asks us to write the results using the principles discussed in the Method of Presentation section.

Results

A Kruskal-Wallace test was applied to the ranked data relating the affective content of words to learning, using an alpha level of .05. The resulting value of H was statistically nonsignificant, H (2, N = 15) = 2.78, ns.

Chapter 17: Two-Way Between-Subjects Analysis of Variance

Study Objectives

This chapter covers the basics of factorial analysis of variance. After reading the material, you should be able to define what a factorial design is and ways of characterizing them (e.g., the difference between a 2x3 factorial design and a 3x3 factorial design). You should be able to specify the conditions when we typically apply factorial analysis of variance. You should be able to define and explain what a main effect is and what an interaction effect is. You should know how to graph an interaction effect. You should be able to describe how the between group variability is partitioned into variability due to each of the main effects and the interaction effect.

You should be able to compute all relevant sum of squares, degrees of freedom, mean squares and F tests for factorial analysis of variance problems. You should be able to form a summary table to show the results of these calculations.

You should be able to conduct an F test to evaluate the viability of all the null hypotheses in a factorial analysis of variance. You should be able to explain the assumptions underlying the F test and briefly characterize the robustness of the tests to assumption violations.

You should be able to compute eta squares for each main effect and interaction effect and be able to interpret them.

You should be able to apply the Tukey HSD test as a follow-up analysis to determine the nature of the relationship for a main effect. You should be able to apply the strategy of interaction comparisons to determine the nature of an interaction effect (using the modified Bonferroni method).

You should be able to discuss the complications that result in factorial analysis of variance when there are unequal sample sizes in each cell and specify how statisticians deal with the complications.

You should be able to determine the sample size necessary to achieve a given level (e.g., 0.80) of statistical power when designing a study that will use factorial analysis of variance. You should be able to explain how alpha, sample size, and the value of eta squared in the population influences statistical power for this test.

You should be able to write up the results of a factorial analysis of variance using APA format as discussed in the Method of Presentation section. You should be able to interpret the results of a factorial analysis of variance from examples reported in the literature.

Study Tips

One of the most difficult concepts for students to grasp is that of an interaction effect. You should make sure that you understand this concept, as it is essential for an understanding of factorial analysis of variance. In addition to the material in the text, some people have characterized interaction effects as *moderated* relationships, and diagramed them, conceptually, as follows:

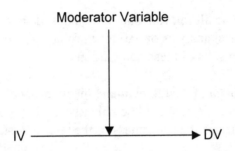

The arrows indicate the hypothesized presence of a causal relationship between variables (e.g., the arrow from the independent variable, IV, and the dependent variable, DV, indicates that the IV causes the DV). In the moderator approach, one of the two factors in the factorial design is designated the independent variable and the other factor is designated the moderator variable. The moderator variable is said to influence either the strength or the nature of the relationship between the independent variable and the dependent variable (hence the arrow going to the arrow). For example, using the example in the textbook, suppose that the independent variable is religion, the dependent variable is the number of children someone wants to have, and the moderator variable is religiosity. In the above diagram, religiosity is said to moderate the impact of religion on the number of preferred children. For example, it may be that there are differences in the preferred number of children between Catholics and Protestants for religious individuals, but not for non-religious individuals. In this case, the effect of religion on the preferred number of children *depends upon* religiosity. The choice of which variable is conceptualized as the moderator variable and which as the primary independent variable does not matter statistically, but it may be important from a theoretical standpoint. In any case, it may help you to think of interaction

208

effects in terms of moderator variables. Study Figures 17.1 and 17.2 in the textbook carefully and make sure you understand why an interaction effect is present or absent in each scenario depicted.

Another common point of confusion is with respect to the follow-up tests to determine the nature of a relationship for a statistically significant main effect. Some students try to apply the Tukey HSD test when the main effect has only two levels. This is not appropriate. The Tukey HSD test is used only when the main effect has three or more levels. When there are just two levels, the nature of the relationship is discerned by simple inspection of the means (similar to what we did with the independent groups t test).

There are many formulas and calculations in this chapter. Again, try not to lose sight of the bigger picture and what you are trying to accomplish with all of the calculations.

Glossary Of Important Terms

Study the terms listed below. Make sure you understand each so that you could explain them to someone else who does not know them.

Two-Way Between-Subjects Analysis of Variance: A statistical method for analyzing the effects of two independent variables, considered jointly, on a dependent variable in the context of a factorial design.

Factors: Another name for an independent variable.

Factorial Design: An experimental design in which all levels of one independent variable occur at all levels of the other independent variable.

Cell: Each unique combination of values of two variables in a factorial design.

Main Effects: In a factorial design with two independent variables, a main effect is the effect of one of the independent variables considered separately. For the main effect of Factor A, we ask, "is Factor A related to the dependent variable" (i.e., are the population means across the levels of factor A the same)?

Interaction Effects: In a factorial design with two independent variables, an interaction effect addresses the question of whether the strength or nature of the relationship between one of the independent variables and the dependent variable differs as a function of the other independent variable.

Simple Main Effects Analysis: An analytic method that can, in some cases, determine the nature of an interaction.

Interaction Comparisons: An analytic method for determining the nature of an interaction effect and which focuses on 2 X 2 subtables within a factorial design.

Modified Bonferroni Procedure: A method for controlling Type I errors across a series of comparisons or tests.

Least Squares Factorial Analysis of Variance: A strategy for analyzing factorial analysis of variance data when there are unequal sample sizes.

Practice Questions: True-False Format

1. The joint effects of two independent variables on a dependent variable can be studied using factorial designs.

2. A factorial design having two factors can be represented by the notation a x b, where a refers to the number of levels of the first factor and b refers to the number of levels of the second factor.

3. The number of groups required in a between-subjects factorial design is simply the sum of the number of levels of each factor.

4. Factorial designs can not be used to examine the relationship between a dependent variable and three or more independent variables.

5. The conditions for using two-way between-subjects analysis of variance are similar to those for the independent groups t test and one-way between-subjects analysis of variance, except that two independent variables rather than one independent variable are studied.

6. The first two questions addressed in a two-way analysis of variance are concerned with interaction effects.

7. Each main effect in a two-way analysis of variance has a null hypothesis and an alternative hypothesis associated with it.

8. Assuming there are unequal numbers of subjects in each group, the main effect means can be calculated by taking the sum of the two cells in a given column.

9. An interaction effect refers to the comparison of cell means in terms of whether the relationship between one of the independent variables and the dependent variable differs as a function of the other independent variable.

10. It is impossible to obtain a statistically significant interaction effect if one of the main effects is statistically significant.

11. Stated informally, the null hypothesis for an interaction effect is that the nature and/or magnitude of the mean differences on the independent variable as a function of the other independent variable differs depending on the level of the dependent variable.

12. When dealing with population means, the determination of whether an interaction is present is most readily made by examining the slopes of the lines in a given graph.

13. When data are presented graphically, the presence of an interaction is suggested by parallel lines.

14. The principles for detecting main effects and interactions are not readily generalizable from 2 x 2 designs to designs with more than two levels for one or both factors.

15. When dealing with sample data, we can determine the existence of main effects and interactions in populations by mere visual inspection of sample means or slopes of lines in a graph.

16. Due to the role of sampling error, nonequivalent sample means do not necessarily indicate nonequivalent population means.

17. The sum of squares between in two-way analysis of variance is partitioned into three components: (1) variability due to the first independent variable, (2) variability due to the second independent variable, and (3) variability due to the interaction of the two independent variables.

18. The total sum of squares in two-way between-subjects analysis of variance can be represented as $SS_{TOTAL} = SS_{WITHIN} + SS_{ERROR}$.

19. In two-way analysis of variance, each of the components of $SS_{BETWEEN}$ is multiplied by its respective degrees of freedom and the resulting mean squares, MS_A, MS_B, and MS_{AxB}, are then divided by MS_{TOTAL} to yield t ratios.

20. The F ratio formed by MS_B/MS_{WITHIN} is used to test the null hypothesis with respect to the main effect of factor B.

21. As with the sums of squares, the mean squares are additive, such that:

$$MS_{TOTAL} = MS_A + MS_B + MS_{AxB} + MS_{WITHIN}$$

22. In two-way between-subjects analysis of variance, $df_{TOTAL} = N - 1$.

23. In two-way between-subjects analysis of variance, the relevant mean squares are obtained by dividing the sum of squares for each source of between-group variability by its degrees of freedom.

24. An eta squared can be computed from the information provided in a summary table.

25. The test of the null hypothesis for the main effect of factor A is made with reference to the F value derived from SS_A/MS_A.

26. The F tests for the three sources of variability in two-way between-subjects analysis of variance are based on the same assumptions that underlie one-way between-subjects analysis of variance.

27. When the sample sizes in each cell are moderate (e.g., greater than 20) to large in size and the same for all groups, the F test is quite robust to even marked departures from normality.

28. As the number of groups in the factorial design increases, violation of the assumption of homogeneity of population variances becomes increasingly problematic.

29. Eta-squared can be used to assess the strength of the relationships for the two main effects, but not for the interaction effect.

30. When a statistically significant main effect has only two levels, the nature of the relationship is determined in the same fashion as for the independent groups t test; if a statistically significant main effect has three or more levels, then the nature of the relationship is determined using an HSD procedure.

31. When an interaction effect is statistically significant, the nature of the interaction can be determined using simple main effects analysis or interaction comparisons.

32. Simple main effects analyses focus on 2 X 2 subtables within a factorial design.

Answers to True-False Items

1. T	11. F	21. F	31. T
2. T	12. T	22. T	32. F
3. F	13. F	23. T	
4. F	14. F	24. T	
5. T	15. F	25. F	
6. F	16. T	26. T	
7. T	17. T	27. T	
8. F	18. F	28. T	
9. T	19. F	29. F	
10. F	20. T	30. T	

Practice Questions: Short Answer

1. Under what conditions is the two-way between-subjects analysis of variance used?

2. Two-way factorial designs allow us to address three important questions. What are they?

3. What form does a main effect take in a two-way analysis of variance?

4. What is an interaction effect?

5. Discuss the partitioning of variability in the context of two-way analysis of variance.

6. What are the assumptions of the F tests in two-way analysis of variance?

7. Discuss the robustness of the F tests in two-way analysis of variance.

8. Discuss the analysis of main effects in terms of the nature of the relationship in two-way analysis of variance.

9. Discuss the analysis of interaction effects in terms of the nature of the relationship in two-way analysis of variance.

10. What problem might arise in using interaction contrasts to determine the nature of an interaction effect?

Answers to Short Answer Questions

1. The conditions for using two-way between-subjects analysis of variance are similar to those for the independent groups t test and one-way between-subjects analysis of variance, except that two independent variables rather than one are studied. Thus, two-way between-subjects analysis of variance is typically applied when (1) the dependent variable is quantitative in nature and is measured on a level that at least approximates interval characteristics; (2) the independent variables are both between-subjects in nature (they can be either qualitative or quantitative); (3) the independent variables both have two or more levels; (4) the independent variables are combined to form a factorial design.

2. Two-way factorial designs allow us to address three issues concerning the relationship between a dependent variable and two independent variables. Specifically: (1) Is there a relationship between independent variable A, considered alone, and the dependent variable? (2) Is there a relationship between independent variable B, considered alone, and the dependent variable? (3) Is there a relationship between independent variable A and independent variable B, considered in combination, and the dependent variable, independent of the effects of
variable A alone and variable B alone?

3. Each main effect in a two-way analysis of variance has a null hypothesis and an alternative hypothesis associated with it. In each instance, the null hypothesis states that the population means for the groups constituting that effect are equal to one another and the alternative hypothesis states that these means are not all equal to one another. These hypotheses take the same form as they would for an independent groups t test (if there are two levels of an independent variable) or a one-way between-subjects analysis of variance (if there are three or more levels).

4. An interaction effect refers to the comparison of cell means in terms of whether the nature of the relationship between one of the independent variables and the dependent variable differs as a function of the other independent variable. Stated differently, an interaction effect refers to the case where the nature or strength of the relationship between one of the

214

independent variables and the dependent variable differs as a function of the other independent variable.

5. As was the case in one-way analysis of variance, the sum of squares total in two-way analysis of variance can be partitioned into a sum of squares between and a sum of squares within:

$$SS_{TOTAL} = SS_{BETWEEN} + SS_{WITHIN}$$

However, the overall sum of squares between is further partitioned into three components: (1) variability due to the first independent variable (designated as "factor A"), (2) variability due to the second independent variable (designated as "factor B"), and (3) variability due to the interaction of factors A and B (symbolized "A x B"). The sum of squares total can thus be represented as

$$SS_{TOTAL} = SS_A + SS_B + SS_{AxB} + SS_{WITHIN}$$

6. The F tests in a two-way between-subjects analysis of variance are based on the same assumptions that underlie one-way between-subjects analysis of variance. These are: (1) the samples are independently and randomly selected from their respective populations; (2) the scores in each population are normally distributed; (3) the scores in each population have homogeneous variances. In addition, the dependent variable should be quantitative in nature and measured on a level that at least approximates interval characteristics.

7. For the F tests to be valid, it is important that the assumption of independent and random selection be met. As with one-way analysis of variance, under certain conditions the F tests are robust to violations of the normality and homogeneity of variance assumptions. This is particularly true when sample (cell) sizes are equal and relatively large. When the sample sizes in each cell are moderate (e.g., greater than 20) to large in size and the same for all groups, the F test is quite robust to even marked departures from normality. As the number of groups in the factorial design increases, heterogeneity of population variances becomes increasingly problematic. Even with a relatively large number of groups, the F test is relatively robust when the population variance of one group is as much as two to three times larger than the population variances for the other groups. however, the Type I error rate begins to become unacceptable when the population variance of one group is four or more times larger than the population variance for the other group. In the case of highly different sample sizes across groups, the robustness of the F test to non-normality and variance heterogeneity often diminishes, sometimes considerably so.

8. When a statistically significant main effect has only two levels, the nature of the relationship is determined in the same fashion as for the independent groups t test. This involves making inferences about the two population means from examination of the two sample means. If a statistically significant main effect has three or more levels, then the nature of the relationship is determined using an HSD procedure. This involves computing the absolute difference between all possible pairs of sample means comprising the main effect and then comparing each of these against a critical difference.

9. When an interaction effect is statistically significant, the nature of the interaction can be determined using a variety of statistical procedures. In general, statisticians recommend an approach called interaction comparisons over other strategies.

Interaction comparisons focus on 2 X 2 subtables within a factorial design. When the overall design is a 2 X 2 design, there is no need to conduct any additional analyses. The nature of the interaction effect is apparent by examination of the cell means. For more complex designs, interaction comparisons involve breaking an overall factorial design into a series of 2 X 2 subtables. Interaction comparisons involve performing a separate 2 X 2 factorial analysis of variance for each of these subtables.

10. A problem with the strategy of breaking an overall factorial design into a series of 2 X 2 subtables and then performing a separate 2 X 2 factorial analysis of variance for each of these subtables is that it does not adequately control the experimentwise error rate at α. One way of controlling this problem is to use a technique called a modified Bonferroni procedure.

Answers to Selected Exercises from Textbook

Exercise Number 49: Exercise 49 asks us to analyze the data as completely as possible and to draw conclusions for each effect. There are two independent variables: weight, which has two levels (overweight and normal) and time cue, which has three levels (11:00, 12:00, and 1:00). Different individuals are used in each cell and the levels of one independent variable are combined with the levels of the other. Thus, we have a 2 X 3 between-subjects factorial design. The dependent variable is the subjects' ratings of how hungry they felt.

Begin the two-way between-subjects analysis of variance by preparing the data for computation of the sums of squares formulas. The data are presented in the computation table below:

Weight (B)

Time (A)	Overweight X X²	Normal X X²	Main Effect of Time Cue
11:00	5 25 6 36 7 49 6 36 6 36 $T_{11O} = 30$ $\overline{X}_{11O} = 6.00$ $T_{11O}^2 = 900$	6 36 6 36 7 49 5 25 6 36 $T_{11N} = 30$ $\overline{X}_{11N} = 6.00$ $T_{11N}^2 = 900$	$T_{11} = 60$ $\overline{X}_{11} = 6.00$ $T_{11}^2 = 3600$
12:00	9 81 7 49 8 64 9 81 10 100 $T_{12O} = 43$ $\overline{X}_{12O} = 8.60$ $T_{12O}^2 = 1849$	6 36 6 36 5 25 7 49 6 36 $T_{12N} = 30$ $\overline{X}_{12N} = 6.00$ $T_{12N}^2 = 900$	$T_{12} = 73$ $\overline{X}_{12} = 7.30$ $T_{12}^2 = 5329$
1:00	10 100 10 100 9 81 8 64 9 81 $T_{1O} = 46$ $\overline{X}_{1O} = 9.20$ $T_{1O}^2 = 2116$	7 49 6 36 7 49 8 64 5 25 $T_{1N} = 33$ $\overline{X}_{1N} = 6.60$ $T_{1N}^2 = 1089$	$T_1 = 79$ $\overline{X}_1 = 7.90$ $T_1^2 = 6241$
Main Effect of Weight	$T_O = 119$ $\overline{X}_O = 7.93$ $T_O^2 = 14161$	$T_N = 93$ $\overline{X}_N = 6.20$ $T_N^2 = 8649$	

217

We can also prepare the various sums we need to compute the sums of squares:

$$\Sigma X = 5 + 6 + ... + 8 + 5 = 212.00$$

$$\Sigma X^2 = 25 + 36 + ... + 64 + 25 = 1,570.00$$

$$\frac{(\Sigma X)^2}{N} = \frac{(212)^2}{30} = 1498.13$$

$$\frac{\Sigma T_{Ai}^2}{nb} = \frac{(60)^2 + (73)^2 + (79)^2}{(5)(2)} = 1517.00$$

$$\frac{\Sigma T_{Bi}^2}{na} = \frac{(119)^2 + (93)^2}{(5)(3)} = 1520.67$$

$$\frac{\Sigma T_{AiBi}^2}{n} = \frac{900 + 900 + 1849 + 900 + 2116 + 1089}{5} = 1550.80$$

Now all we need to do is substitute the appropriate values into the sum of squares formulas:

$$SS_{TOTAL} = \Sigma X^2 - \frac{(\Sigma X)^2}{N}$$

$$= 1570 - 1498.133 = 71.867$$

$$SS_A = \frac{\Sigma T_{Ai}^2}{nb} - \frac{(\Sigma X)^2}{N}$$

$$= 1517 - 1498.133 = 18.867$$

$$SS_B = \frac{\Sigma T_{Bi}^2}{na} - \frac{(\Sigma X)^2}{N}$$

$$= 1520.667 - 1498.133 = 22.534$$

$$SS_{WITHIN} = \Sigma X^2 - \frac{\Sigma T_{AiBi}^2}{n}$$

$$= 1570 - 1550.8 = 19.2$$

$$SS_{A \times B} = \frac{(\Sigma X)^2}{N} + \frac{\Sigma T_{AiBi}^2}{n} - \frac{\Sigma T_{Ai}^2}{nb} - \frac{\Sigma T_{Bi}^2}{na}$$

$$= 1498.133 + 1550.8 - 1520.667 - 1517 = 11.226$$

Now we compute the degrees of freedom.

$$df_A = a - 1 = 3 - 1 = 2$$

$$df_B = b - 1 = 2 - 1 = 1$$

$$df_{A \times B} = (a - 1)(b - 1) = (3 - 1)(2 - 1) = 2$$

$$df_{WITHIN} = (a)(b)(n-1) = (3)(2)(5-1) = 24$$

$$df_{TOTAL} = N - 1 = 30 - 1 = 29$$

The mean squares are obtained by dividing the sum of sqaures by its corresponding degrees of freedom:

$$MS_A = \frac{SS_A}{df_A} = \frac{18.867}{2} = 9.4335$$

$$MS_B = \frac{SS_B}{df_B} = \frac{22.534}{1} = 22.534$$

$$MS_{A \times B} = \frac{SS_{A \times B}}{df_{A \times B}} = \frac{11.266}{2} = 5.633$$

$$MS_{WITHIN} = \frac{SS_{WITHIN}}{df_{WITHIN}} = \frac{19.2}{24} = .80$$

The F ratios are next:

$$F = \frac{MS_A}{MS_{WITHIN}} = \frac{9.4335}{.80} = 11.792$$

$$F = \frac{MS_B}{MS_{WITHIN}} = \frac{22.534}{.80} = 28.168$$

$$F = \frac{MS_{A \times B}}{MS_{WITHIN}} = \frac{5.633}{.80} = 7.041$$

Finally, we can summarize our computations in a summary table:

Source	SS	df	MS	F
A (Time Cue)	18.867	2	9.434	11.792
B (Weight)	22.534	1	22.534	28.168
A x B	11.266	2	5.633	7.041
Within	19.200	24	.800	
Total	71.867	29		

The null hypothesis for the main effect of weight states that the population means for the two groups are equal. The alternative hypothesis states that the population means are not equal.

$$H_0: \mu_O = \mu_N$$

$$H_1: \mu_O \neq \mu_N$$

The critical value of F from Appendix F for an alpha level of .05 and 2 and 24 degrees of freedom is 3.40. Because the observed F of 11.792 for weight is greater than 3.40, we reject the null hypothesis and conclude that there is a relationship between weight and how hungry an individual felt.

The null hypothesis for the main effect of time cue states that the population means for the three groups are equal. The alternative hypothesis states that the population means are not equal.

$$H_0: \mu_{11} = \mu_{12} = \mu_1$$

$$H_1: \text{the three population means are not all equal}$$

The critical value of F from Appendix F for an alpha level of .05 and 1 and 24 degrees of freedom is 4.26. Because the observed F of 28.168 for weight is greater than 4.26, we reject the null hypothesis and conclude that there is a relationship between the time manipulation and how hungry subjects' felt.

The null hypothesis for the interaction effect states that the relationship between time cues and hunger is the same for overweight and normal weight subjects. The alternative hypothesis states that the relationship between time cues and hunger depends on the subject's weight. The critical value of F from Appendix F for an alpha level of .05 and 2 and 24 degrees of freedom is 3.40. Because the observed F of 7.041 is greater than 3.40, we reject the null hypothesis and conclude that an interaction effect exists.

The strength of the relationship is analyzed using eta-squared:

$$eta^2_A = \frac{SS_A}{SS_{TOTAL}} = \frac{18.867}{71.867} = .263$$

$$eta^2_B = \frac{SS_B}{SS_{TOTAL}} = \frac{22.534}{71.867} = .314$$

$$eta^2_{A \times B} = \frac{SS_{A \times B}}{SS_{TOTAL}} = \frac{11.266}{71.867} = .157$$

The strongest effect is for the main effect of time cue. The proportion of variability in subjects' ratings of how hungry they felt that is associated with time is .314. This represents a strong effect. The proportion of variability in subjects' ratings of how hungry they felt that is associated with weight is .263. This represents a strong effect. Although the interaction is statistically significant, only .157 or 16% of the variability in peoples' ratings of how hungry they felt is associated with the joint effects of the independent variables. This represents a moderate effect.

Since the independent variable "weight" has only two levels, the nature of the relationship between weight and hunger is determined through examination of the sample means. For this variable, the mean for the overweight group (7.93) is greater than the mean for the normal group (6.20), indicating that the overweight group felt hungrier than the normal group.

Because the independent variable "time cue" has three levels, the nature of the relationship between the time cue and hunger is determined using the Tukey HSD procedure. The formula for the critical difference for factor A is as follows (Equation 17.25):

$$CD = q \sqrt{\frac{MS_{WITHIN}}{nb}}$$

We obtain the value of q from Appendix G. For an overall alpha level of .05, degrees of freedom within of 24, and three levels of factor A, the value of q from Appendix G is 3.53.

$$CD = 3.53 \sqrt{\frac{.80}{(5)(2)}} = .998$$

The critical difference is thus .998. Now we can finish up the HSD analysis by comparing the difference between all possible combinations of two group means to the CD value:

Null Hypothesis tested	Absolute difference between sample means	Value of CD	Null Hypothesis rejected?
$\mu_{11am} = \mu_{12pm}$	$\lvert 6.00 - 7.30 \rvert = 1.30$	.998	Yes
$\mu_{11am} = \mu_{1pm}$	$\lvert 6.00 - 7.90 \rvert = 1.90$	.998	Yes
$\mu_{12pm} = \mu_{1pm}$	$\lvert 7.30 - 7.90 \rvert = 0.60$	.998	No

The results indicate that the nature of the relationship between time cue and hunger is such that the 11 a.m. cue produced significantly lower hunger ratings than the 12 or 1 p.m. cues. The 12 and 1 p.m. cues did not differ from each other.

Because the interaction effect was statistically significant, we use interaction comparisons to explore the nature of the interaction. There are three 2 X 2 sub-tables.

	Overweight	Normal
11 A.M.	6.00	6.00
12 P.M.	8.60	6.00

	Overweight	Normal
11 A.M.	6.00	6.00
1 P.M.	9.20	6.60

	Overweight	Normal
12 P.M.	8.60	6.00
1 P.M.	9.20	6.60

Using equation 17.27, we calculate the sum of squares for each of the interaction effects in the 2 X 2 sub-tables. For the first table, the sum of squares is:

$$SS_{AxB(1)} = \frac{[(n\,X_a + n\,X_d) - (n\,X_b + n\,X_c)]^2}{4\,n}$$

$$= \frac{[(5\,(6.00) + 5\,(6.00)) - (5\,(6.00) + 5\,(8.60))]^2}{4\,(5)}$$

$$= \frac{(60 - 73)^2}{20} = 8.45$$

For the second sub-table, the sum of squares is:

$$SS_{AxB(2)} = \frac{[(5\,(6.00) + 5\,(6.60)\,) - (5\,(6.00) + 5\,(9.20)\,)\,]^2}{4\,(5)}$$

$$= \frac{(63 - 76)^2}{20} = 8.45$$

For the third sub-table, the sum of squares is:

$$SS_{AxB(3)} = \frac{[(5\,(8.60) + 5\,(6.60)\,) - (5\,(6.00) + 5\,(9.20)\,)\,]^2}{4\,(5)}$$

$$= \frac{(76 - 76)^2}{20} = 0.00$$

We then form an F ratio for each subtable, using MS_{WITHIN} from the overall summary table:

$$F_{AxB(1)} = \frac{8.45}{.80} = 10.563$$

$$F_{AxB(2)} = \frac{8.45}{.80} = 10.563$$

$$F_{AxB(3)} = \frac{0.00}{.80} = 0$$

We next apply the modified Bonferroni procedure to control for experimentwise error rate. We determine an exact p value for each of the above F ratios (based on 1 and 24 degrees of freedom). This usually must be done with the aid of a computer. The p values are ordered in a column from smallest to largest and each p value is compared against a "critical alpha." If the p value is less than the "critical alpha," then the interaction comparison is declared statistically significant. Here is a table that contains the relevant information:

Contrast	F ratio	p value	Critical alpha
$AXB_{(1)}$	10.563	.0034	.05/3 = .0167
$AXB_{(2)}$	10.563	.0034	.05/2 = .025
$AXB_{(3)}$	0.00	ns	.05/1 = .050

The value of the "critical alpha" changes for each successive p value going from the top row to the bottom row of the table. If we use the traditional α of .05 for our per comparison error rate, then the largest F ratio must yield a p value less than α divided by the total number of contrasts for the null hypothesis to be rejected. The next largest F ratio must yield a p value less than α divided by the number of contrasts minus one for the null hypothesis to be rejected. The next largest F ratio must yield a p value less than α divided by the number of contrasts minus two for the null hypothesis to be rejected. And so on, subtracting an additional unit from the total number of contrasts with each successive contrast. As soon as the first statistically nonsignificant contrast is encountered, as one proceeds from the top row to the bottom row, all remaining contrasts are declared statistically nonsignificant.

In this example, two of the interactions yielded identical p values, so it is ambiguous which should be declared to have the "smallest p value" for purposes of ordering them for the modified Bonferroni test. However, in this case, the issue is moot, because both interactions will be declared statistically significant no matter which is ordered first. Another unusual

226

feature is an F ratio of zero for the third interaction. For reasons we will not explore here, this is a highly unusual event and actually yields an infinitely small p value. However, when an F ratio takes on values much less than 1.0, they are declared statistically non-significant, without further analysis.

In our analysis, the first two sub-tables yielded statistically significant interaction effects. This leads us to the following statements about the interaction: The effect of time cue on hunger differs as a function of the subject's weight. Among overweight subjects, the 12 p.m. cue produced greater feelings of hunger than the 11 a.m. cue (8.60 - 6.00 = 2.60), an effect that was not observed among individuals of normal weight (6.00 - 6.00 = 0.0). Similarly, among overweight subjects, the 1 p.m. cue produced greater feelings of hunger than the 11 a.m. cue (9.20 - 6.00 = 3.20), but this effect is diminished among individuals of normal weight (6.60 - 6.00 = 0.60). Finally, the difference between the 12 p.m. and the 1 p.m. cues was the same for overweight (9.20 - 8.60 = .60) and normal weight (6.60 - 6.00 = .60) individuals.

The write-up following the principles discussed in the Method of Presentation section appears as follows (Note: All standard deviations were computed using Equation 7.2 for a standard deviation estimate in Chapter 7) :

Results

Hunger ratings were subjected to a two-way analysis of variance having three levels of external cues (time reported as 11 a.m., 12 p.m., and 1 p.m.) and two levels of weight overweight and normal). All tests used an alpha level of .05. The main effect of external cue was statistically significant (F (2,24) = 11.79, p < .01), with an eta squared value of .26. The nature of the effect was determined using the Tukey HSD test. Results showed that an 11 a.m. cue (M = 6.00, SD = .67) produced significantly lower hunger ratings than the 12 p.m. (M

= 7.30, $\underline{SD}$ = 1.64) or 1 p.m. cues ($\underline{M}$ = 7.90, $\underline{SD}$ = 1.66). There was not a significant difference, however, between the 12 p.m. and 1 p.m. cues. The main effect of weight showed that overweight subjects ($\underline{M}$ = 7.93, $\underline{SD}$ = 1.67) tended to rate themselves as hungrier than normal weight subjects ($\underline{M}$ = 6.20, $\underline{SD}$ = .86), $\underline{F}$(1, 24) = 28.17, $\underline{p}$ < .01. The strength of the relationship, as indexed by eta-squared, was .31.

The interaction effect was statistically significant ($\underline{F}$(2, 24) = 7.04, $\underline{p}$ < .01), and yielded an eta squared of .16. The nature of the interaction was analyzed using interaction comparisons coupled with a modified Bonferroni procedure to maintain the experimentwise error rate at .05. The results showed that the effect of time cue on hunger changed as a function of the individual's weight. Among overweight individuals, the difference in mean hunger ratings between the 12 p.m. and 11 a.m. cues (Means = 8.60 - 6.00 = 2.60) was statistically significantly (p < .05) larger than the difference between mean hunger ratings for these cues for normal weight individuals (Means = 6.00 - 6.00 = 0). Similarly, for overweight individuals, the difference in mean hunger ratings between the 1 p.m. cue and the 11 a.m. cue (9.20 - 6.00 = 3.20) was statistically significantly (p < .05) larger than for normal weight individuals (Means = 6.60 - 6.00 = 0.60). Finally, the difference between the 12 p.m. and the 1 p.m. cues for overweight as compared to normal weight individuals was not statistically significant (Means = 8.60 - 9.20 versus 6.60 - 6.00, respectively).

Chapter 18: Overview and Extension: Selecting the Appropriate Statistical Test for Analyzing Bivariate Relationships and Procedures for More Complex Designs

Study Objectives

This chapter discusses issues to consider when selecting a statistical test for data analysis. It also introduces more advanced statistical methods that you may encounter in the research literature. After reading the material in this chapter, you should be able to state the three questions that have guided our analysis of bivariate relationships. You should be able to identify the statistical procedure designed to address each question for each of the statistical tests we have considered. You should be able to identify the four cases that involve quantitative and qualitative independent and dependent variables that dictate the type of analysis pursued. You should be able to describe the statistical tests that are appropriate in each case.

You should be able to characterize multivariate statistics in general, as well as the following methods: two way repeated measures analysis of variance; between-within analysis of variance, multivariate analysis of variance; Hotelling's T^2 approach; multiple regression (including a multiple correlation and a regression coefficient); logistic regression; factor analysis; and log linear analysis.

Study Tips

If you master the material in this chapter, you should be able to choose how to analyze data for a given data analytic problem. Try going to the library and getting some social science journals that report empirical data (e.g., the *Journal of Experimental Psychology*, the *Journal of Experimental Social Psychology*, the *Journal of Applied Psychology*, the *Journal of Applied Social Psychology*, the *Journal of Abnormal Psychology*). Read some of the research reports and identify the method of analysis that the researchers used. Specify the criteria that probably influenced the authors' choice of analytic method, as described in this chapter.

Try designing some experiments of your own. After specifying the rationale, hypotheses and design, specify what methods of data analysis you would use and why. Discuss your decisions with your instructor during his or her office hours.

After reading this chapter, you should appreciate the importance of identifying the independent and dependent variables of a study, whether they are qualitative or quantitative, whether the independent variable is between-subjects or within subjects, and considering the number of values of the independent and dependent variables. Make sure you can read a research report or an experimental description and distinguish these features. Review all of the "Exercises to Apply Concepts" throughout each chapter of the book and reaffirm the criteria for each example described.

Glossary Of Important Terms

Study the terms listed below. Make sure you understand each so that you could explain them to someone else who does not know them.

Multivariate Statistics: Statistical techniques for analyzing relationships when there are more than one dependent variable and/or when there are two or more independent variables.

Two-Way Repeated Measures Analysis of Variance: A statistical method for analyzing data for a two factor design when both independent variables are within-subjects in nature.

Two-Way Between-Within Analysis of Variance: A statistical method for analyzing data for a two factor design when one independent variable is within-subjects in nature and the other is between-subjects in nature.

Multivariate Analysis of Variance: A statistical method for analyzing one-way designs and factorial designs when there is more than one dependent variable.

Hotelling T test: The multivariate extension of the independent groups t test. It is used under the same conditions as the independent groups t test, except there is more than one dependent variable.

Multiple Regression: A statistical technique that extends bivariate linear regression to the case where there is more than one independent or predictor variable.

Regression Coefficient: A parameter in a multiple regression equation that is analogous to a slope. It reflects the number of units the criterion variable is predicted to change for each unit change in a given predictor variable, when the effects of the other predictor variables are held constant.

Squared Multiple Correlation Coefficient: An index of the strength of the relationship between the criterion variable and the set of predictor variables in a multiple regression analysis.

Factor Analysis: A statistical technique used to determine if the correlations among a set of variables can be accounted for by one or more underlying dimensions, or factors.

Log-linear Analysis: A statistical method that is conceptually similar to chi-square analysis and which is usually used in cases where there is a qualitative dependent variable and more than one qualitative independent variables. Log-linear analysis possesses statistical properties that make it suitable for the simultaneous analysis of multiple between-subjects qualitative variables.

Practice Questions: True-False Format

1. When both variables under study are qualitative and between-subjects in nature, the appropriate method of data analysis is the chi square test as discussed in Chapter 15.

2. If both variables are qualitative, but one or both are within-subjects in nature, the appropriate method of data analysis is the chi square test as discussed in Chapter 15..

3. The major factor to be weighed when deciding whether to use a parametric or a nonparametric test is the decision about what aspect of the distribution that you want to compare groups on.

4. If means (or linear relationships) are the focus, then a nonparametric test is the analysis of choice.

5. If ranks are the focus, then parametric methods should be pursued.

6. The choice of which distributional parameter to compare across groups is often dictated by theory or the substantive question of interest.

7. The observed distribution may bias one's decision about a parameter to focus on.

8. An investigator should never focus statistical tests on multiple parameters (means, medians, ranks).

9. In situations where the dependent measure has only two or three values, most researchers will not use a parametric test.

10. A dependent variable can still be normally distributed, even if it has only two or three values and, hence, the assumption of nonparametric tests will be violated.

11. With only two or three values on the dependent variable, there will be a large number of ties in ranks, and numerous ties can create problems for ranked-based approaches.

12. If the dependent variable is measured on an ordinal level that seriously departs from interval level characteristics, a parametric test is always the optimal test.

13. A non-parametric test should be pursued if there is reason to believe that the distributional assumptions of the corresponding parametric test have been violated to the extent that the parametric test is not robust.

14. Generally speaking, parametric methods are not conducive to analyzing qualitative variables.

15. The appropriate statistical methods for Case III situations are techniques called logistic regression and polychotomous logistic regression.

16. Given two between-subject variables that are measured on a level that at least approximates interval characteristics, the most common method of analysis is the chi square test.

17. When one or both variables are measured on an ordinal level that seriously departs from interval level characteristics, Pearson correlation is the test of choice.

18. The use of Case II statistics is appropriate when a qualitative variable has many values and there are insufficient number of cases within each level to justify the analysis of means, medians, and the like.

19. Very few research problems require that more than two variables be simultaneously studied.

20. Statistical techniques that analyze the relationship between three or more variables are referred to as multivariate statistics.

21. One example of a multivariate statistic is a factor analysis.

22. Although the joint influence of two between-subjects independent variables on a dependent variable can be studied using two-way between-subjects analysis of variance, it is not possible to study the joint influence of two within-subjects independent variables.

23. When both independent variables are within-subjects in nature, the data can be analyzed using a two-way repeated measures analysis of variance.

24. Sometimes research designs involve one between-subjects independent variable and one within-subjects independent variable.

25. Analysis of variance is rarely used in the behavioral sciences.

26. In situations involving the analysis of two or more dependent variables in a factorial design, analysts will often perform a multivariate analysis of variance.

27. Multivariate analysis of variance tests whether the groups have different population means on each of the dependent variables considered separately.

28. Multiple regression involves the prediction of a criterion variable from two or more predictor variables.

29. In the context of multiple regression, an index of the strength of the relationship between the criterion variable and the set of predictor variables is the regression coefficient.

30. The goal of factor analysis is to determine if the correlations among a set of variables can be accounted for by one or more underlying dimensions, or factors.

Answers to True-False Items

1. T	11. T	21. T
2. F	12. F	22. F
3. T	13. T	23. T
4. F	14. T	24. T
5. F	15. T	25. F
6. T	16. F	26. T
7. T	17. F	27. F
8. F	18. F	28. T
9. T	19. F	29. F
10. F	20. T	30. T

Practice Questions: Short Answer

1. What three questions have guided our study of bivariate relationships?

2. What decisions must be made in selecting a statistical test for situations where the independent variable is qualitative and the dependent variable is quantitative?

3. What factors are involved in deciding whether to use a parametric or a nonparametric test?

4. What is the appropriate statistical test for a dependent variable that has only two or three values?

5. How do we determine if a parametric test is appropriate for a quantitative dependent variable?

6. What are multivariate statistics?

7. What is a multivariate analysis of variance?

8. What is multiple regression?

9. What is factor analysis?

10. What is log-linear analysis?

Answers to Short Answer Questions

1. Three questions have guided our study of bivariate relationships: (1) Given sample data, can we infer that a relationship exists between two variables in the population? (2) If so, what is the strength of the relationship? (3) If so, what is the nature of the relationship?

2. In situations where the independent variable is qualitative and the dependent variable is quantitative, the three decision areas are (1) use of a parametric versus a nonparametric test, (2) whether the independent variable is between-subjects or within-subjects in nature, and (3) the number of levels characterizing the independent variable.

3. The major factor to be weighed when deciding whether to use a parametric or a nonparametric test is the decision about what aspect of the distribution that you want to

compare groups on (e.g., means, ranks). If means (or linear relationships) are the focus, then a parametric test is the analysis of choice. If ranks are the chosen focus, then nonparametric methods should be pursued.

4. In situations where the dependent measure has only two or three values, most researchers will not use a parametric test. As noted in previous chapters, an assumption of parametric tests is that population scores are normally distributed. A dependent variable cannot be normally distributed if it has only two or three values and, hence, the assumption of parametric tests will be violated. Under this circumstance, nonparametric tests will typically be used.

5. Given a quantitative dependent variable with a sufficient range of values, additional characteristics of the data must be considered when determining if a parametric test is appropriate. For example, if the dependent variable is measured on an ordinal level that seriously departs from interval level characteristics, then a nonparametric test might be pursued instead of a parametric test. A nonparametric test should also be pursued if there is reason to believe that the distributional assumptions of the corresponding parametric test have been violated to the extent that the parametric test is not robust.

6. Many research problems require that more than two variables be simultaneously studied. Statistical techniques have been developed to analyze the relationship between three or more variables. Because these techniques consider the variation among multiple variables, they are referred to as multivariate statistics.

7.. Multivariate analysis of variance (MANOVA) is used for multiple dependent variables. It tests whether the groups defined by the independent variable(s) have different population means on the dependent variables considered jointly.

8. Multiple regression involves the prediction of a criterion variable from two or more predictor variables using an extension of the linear model. For example, if we had two predictor variables, X and Z, the linear model would appear as follows:

$$Y = \alpha + \beta_1 X + \beta_2 Z + \epsilon$$

Each predictor has a "slope" (β) that is called a regression coefficient. These coefficients represent the number of units the criterion variable is predicted to change for each unit change in a given predictor variable when the effects of the other predictor variables are held constant.

9. The goal of factor analysis is to determine if the correlations among a set of variables can be accounted for by one or more underlying dimensions, or factors. Factor analysis analyzes the patterning of correlations (or covariances) between all possible pairs of variables in the data set and provides information on the type of factor structure (that is, the number and makeup of factors) that might underlie the data.

10. Sometimes a research question requires that three or more qualitative variables be simultaneously examined. For three such variables, it would be possible to form a three-way contingency table of the observed frequencies. Although the chi-square test can be extended to multidimensional tables of this type, a statistical technique known as log-linear analysis will usually be applied instead. While conceptually similar to chi-square analysis, log-linear analysis possesses statistical properties that make it more suitable for the simultaneous analysis of multiple between-subjects qualitative variables.